Bacteriology

Bacteriology

Haris Russell

R CALLISTO REFERENCE

www.callistoreference.com

Callisto Reference,
118-35 Queens Blvd., Suite 400,
Forest Hills, NY 11375, USA

Visit us on the World Wide Web at:
www.callistoreference.com

ISBN: 978-1-64116-232-6 (Hardback)

Cataloging-in-Publication Data

Bacteriology / Haris Russell.
  p. cm.
Includes bibliographical references and index.
ISBN 978-1-64116-232-6
1. Bacteriology. 2. Microbiology. I. Russell, Haris.
QR41.2 .B33 2019
579--dc23

Table of Contents

Permissions

Index

Preface

Bacteria constitute a large category of prokaryotic microorganisms which are a few micrometers in size. The study of bacteria, their biochemistry, morphology, ecology and genetics is under the scope of bacteriology. The importance of bacteriology is witnessed in its wide applications in the treatment and prevention of diseases with the use of vaccines. It is an important domain under microbiology, which studies the classification, identification and characterization of bacterial species. This textbook aims to shed light on some of the unexplored aspects of bacteriology. It elucidates new techniques and applications of this discipline in a multidisciplinary approach. In this book, constant effort has been made to make the understanding of the difficult concepts of bacteriology as easy and informative as possible, for the readers.

A detailed account of the significant topics covered in this book is provided below:

Chapter 1- Microbiology is the science concerned with the study of unicellular, multicellular or acellular microorganisms. An important sub-discipline within this domain is bacteriology, which studies the biochemistry, morphology, genetics and ecology of bacteria. The aim of this chapter is to provide an introduction to bacteriology.

Chapter 2- The bacterial cell is encapsulated inside a cell membrane that is composed of phospholipids and acts as a barrier to contain proteins, nutrients and other essential components of the cytoplasm. The topics elucidated in this chapter cover some of the important components in the cell structure of bacteria, such as the flagella, plasmid ribosome, mesosomes, gas vacuoles, etc.

Chapter 3- Bacteria exhibit a wide range of metabolic types. These maybe based on the source of energy, source of carbon used for their growth or the electron donor used. An elaborate study of the fundamentals of bacterial metabolism has been provided in this chapter, which includes topics related to bacterial protein, anabolism, catabolism, oxidative phosphorylation etc.

Chapter 4- Bacteria have a single chromosome, plasmids, extra-chromosomal DNAs that consist of genes that are responsible for metabolism, antibiotic resistance and virulence. Bacterial genomes encode between a hundred to a thousand genes. Bacteria grow to maturity and reproduce through binary fission. This chapter discusses the genetics and growth in bacteria through an analysis of bacterial conjugation, measurement of bacterial growth, growth curve, etc.

Chapter 5- Bacteria share several common characteristics, such as unicellularity, lack of nuclear membrane, etc. However, there exist differences in terms of morphology, metabolism, phylogeny, pathogenicity, etc. thus allowing their identification and classification. This chapter covers all the significant classifications of bacteria and also discusses the classification on the basis of flagella, spore, gaseous requirement, etc.

Chapter 6- Bacteria form complex associations with other organisms in the form of symbiotic associations such as mutualism, parasitism and commensalism. This chapter explores the interactions of bacteria with the environment and with other organisms, through a discussion on host-pathogen interaction, transmission methods, bacterial infection methods, etc.

Chapter 7- The diverse uses on bacteria lie in the production of foods, such as cheese, yoghurt and vinegar, production of drugs and vitamins, in genetic engineering and biotechnology, agriculture, biological control of pests, etc. The diverse uses of bacteria in food processing, genetic engineering, pest control, etc. have been covered in this chapter.

Chapter 8- Bacteria can form a parasitic association with other organisms that can lead to disease and death. In humans, these are manifested as various infections such as typhoid fever, tetanus, syphilis, leprosy, etc. The aim of this chapter is to provide an overview of bacterial diseases such as cellulitis, folliculitis, impetigo, vibriosis, gonorrhea, etc.

It gives me an immense pleasure to thank our entire team for their efforts. Finally in the end, I would like to thank my family and colleagues who have been a great source of inspiration and support.

Haris Russell

Chapter 1

Introduction to Bacteriology

Microbiology is the science concerned with the study of unicellular, multicellular or acellular microorganisms. An important sub-discipline within this domain is bacteriology, which studies the biochemistry, morphology, genetics and ecology of bacteria. The aim of this chapter is to provide an introduction to bacteriology.

Microbiology

Microbiology is the study of microorganisms (also known as microbes), which are unicellular or cell-cluster organisms and infectious agents too small to be seen with the naked eye. This includes eukaryotes (organisms with a nucleus), such as fungi and protists, and prokaryotes (organisms without a nucleus), such as bacteria.

A fundamental understanding of how a cell works has come through the study of microorganisms. But microbiology also is an applied science, helping agriculture, health and medicine and maintenance of the environment, as well as the biotechnology industry. Microbiologists study microbes at the level of the community (ecology and epidemiology), at the level of the cell (cell biology and physiology) and at the level of proteins and genes (molecular biology).

Microorganisms are extremely important in our everyday lives. Some are responsible for a significant proportion of the diseases affecting not only humans, but also plants and animals, while others are vitally important in the maintenance and modification of our environment. Still others play an essential role in industry, where their unique properties have been harnessed in the production of food, beverages and antibiotics. Scientists also have learned how to exploit microorganisms in the field of molecular biology, which makes an enormous impact both industrially and medically. Microbiology also encompasses immunology, the study of the body's ability to mount defenses against infectious microbes.

Because microbiology, by definition, studies organisms not visible to the naked eye, we can consider late-17th-century Dutch scientist Antony van Leeuwenhoek the father of the discipline. Leeuwenhoek was the first person to describe tiny cells and bacteria, and he invented new methods for grinding and polishing microscope lenses that allowed for curvatures providing magnifications of up to 270 diameters, the best available lenses at that time. But while van Leeuwenhoek is cited as the first microbiologist, the first recorded microbiological observation — the fruiting bodies of molds — was made earlier, in 1665, by English physicist Robert Hooke.

Other notable people in the history of science who made fundamental discoveries about microorganisms are 19th-century scientists Louis Pasteur and Robert Koch, who are considered the founders of medical microbiology. Pasteur is most famous for his series of experiments designed to disprove the then-widely held theory of spontaneous generation, which solidified microbiology's identity as a biological science.

Pasteur also designed methods for food preservation (pasteurization) and vaccines against several diseases, such as anthrax, fowl cholera and rabies. Koch is best known for his contributions to the germ theory of disease, proving that specific diseases were caused by specific pathogenic microorganisms. He developed a series of criteria that have become known as Koch's postulates. Koch was one of the first scientists to focus on the isolation of bacteria in pure culture, resulting in his description of several novel bacteria, including Mycobacterium tuberculosis, the causative agent of tuberculosis.

Finally, some of the most important discoveries affecting public health occurred in the 20th century, such as the discovery of penicillin by Alexander Fleming, which started a rush to find other natural, and eventually synthetic, antibiotics; the development of vital vaccines, including those for polio and yellow fever; and the birth of molecular biology, which happened in the 1940s with the study of bacteria.

Subdisciplines of Microbiology

Bacteriology

This is the study of bacteria.

Environmental Microbiology

This is the study of the function and diversity of microbes in their natural environments.

Evolutionary Microbiology

This is the study of the evolution of microbes.

Food Microbiology

This is the study of microorganisms causing food spoilage as well as those involved in creating foods such as cheese and beer.

Industrial Microbiology

This is the exploitation of microbes for use in industrial processes, such as industrial fermentation and wastewater treatment. This subdiscipline is linked closely to the biotechnology industry.

Medical (or Clinical) Microbiology

This is the study of the role of microbes in human illness. It includes the study of microbial pathogenesis and epidemiology and is related to the study of disease pathology and immunology.

Microbial Genetics

This is the study of how genes are organized and regulated in microbes in relation to their cellular functions. This subdiscipline is related closely to the field of molecular biology.

Microbial Physiology

This is the study of how the microbial cell functions biochemically. It includes the study of microbial growth, microbial metabolism and microbial cell structure.

Mycology

This is the study of fungi.

Veterinary Microbiology

This is the study of the role in microbes in veterinary medicine.

Virology

This is the study of viruses.

Bacteria

Bacteria are single celled microbes. The cell structure is simpler than that of other organisms as there is no nucleus or membrane bound organelles. Instead their control centre containing the genetic information is contained in a single loop of DNA. Some bacteria have an extra circle of genetic material called a plasmid. The plasmid often contains genes that give the bacterium some advantage over other bacteria. For example it may contain a gene that makes the bacterium resistant to a certain antibiotic.

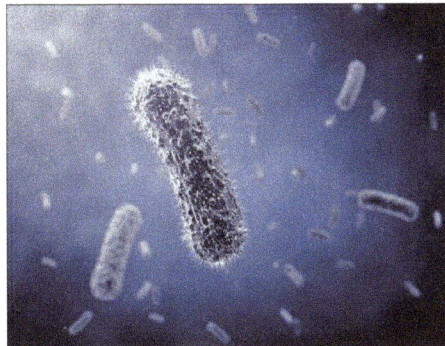

Bacteria are prokaryotes, which consist of a single cell with a simple internal structure.

Bacteria are classified into 5 groups according to their basic shapes: spherical (cocci), rod (bacilli), spiral (spirilla), comma (vibrios) or corkscrew (spirochaetes). They can exist as single cells, in pairs, chains or clusters.

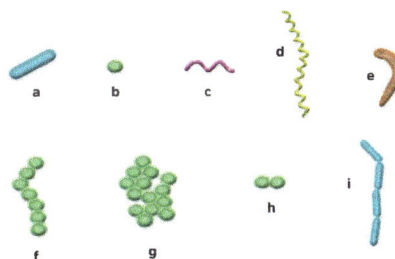

The different bacterial shapes:

a. bacillus (rod), b. coccus (spherical), c. spirillum (spiral), d. spirochaete (corkscrew), e. vibrios (comma), f. chain of cocci, g. cluster of cocci, h. pair of cocci, i. chain of bacilli.

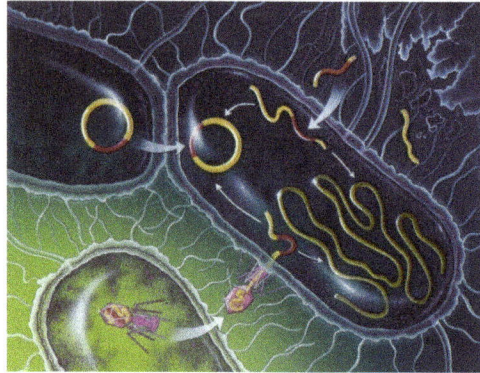

Antibiotic Resistance

Artwork of bacterial cells becoming resistant to antibiotics. This resistance is acquired from a donor cell's plasmid (circular unit of deoxyribonucleic acid, DNA), which has resistance seen at upper left (red/yellow, red is resistance). Viral transmission involves a virus (pink, lower left) obtaining a resistant gene, and passing it to a bacterial cell that incorporates it into its plasmid. Bacterial cells also acquire segments of DNA released from dead cells (upper left). Mutations may also occur, which may be antibiotic resistant and thus allow the bacteria to survive and reproduce.

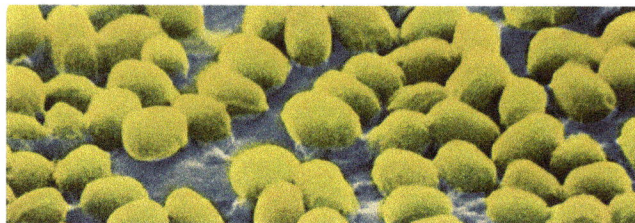

Bacillus Anthracis Spores

This bacterium causes anthrax in farm animals and less commonly in humans. Spores can survive for many years and are resistant to extremes of heat, cold and drying.

Structure

Bacterial cells are different from plant and animal cells. Bacteria are prokaryotes, which means they have no nucleus.

A bacterial cell includes:

- Capsule: A layer found on the outside of the cell wall in some bacteria.
- Cell wall: A layer that is made of a polymer called peptidoglycan. The cell wall gives the bacteria its shape. It is located outside the plasma membrane. The cell wall is thicker in some bacteria, called Gram positive bacteria.

- Plasma membrane: Found within the cell wall, this generates energy and transports chemicals. The membrane is permeable, which means that substances can pass through it.

- Cytoplasm: A gelatinous substance inside the plasma membrane that contains genetic material and ribosomes.

- DNA: This contains all the genetic instructions used in the development and function of the bacterium. It is located inside the cytoplasm.

- Ribosomes: This is where proteins are made, or synthesized. Ribosomes are complex particles made up of RNA-rich granules.

- Flagellum: This is used for movement, to propel some types of bacteria. There are some bacteria that can have more than one.

- Pili: These hair-like appendages on the outside of the cell allow it to stick to surfaces and transfer genetic material to other cells. This can contribute to the spread of illness in humans.

Feeding

Bacteria feed in different ways.

Heterotrophic bacteria, or heterotrophs, get their energy through consuming organic carbon. Most absorb dead organic material, such as decomposing flesh. Some of these parasitic bacteria kill their host, while others help them.

Autotrophic bacteria (or just autotrophs) make their own food, either through either:

- photosynthesis, using sunlight, water and carbon dioxide, or

- chemosynthesis, using carbon dioxide, water, and chemicals such as ammonia, nitrogen, sulfur, and others.

Bacteria that use photosynthesis are called photoautotrophs. Some types, for example, cyanobacteria produce oxygen. These probably played a vital role in creating the oxygen in the earth's atmosphere. Others, such as heliobacteria, do not produce oxygen.

Those that use chemosynthesis are known as chemoautotrophs. These bacteria are commonly found in ocean vents and in the roots of legumes, such as alfalfa, clover, peas, beans, lentils, and peanuts.

Places where Bacteria live

Bacteria can be found in soil, water, plants, animals, radioactive waste, deep in the earth's crust, arctic ice and glaciers, and hot springs. There are bacteria in the stratosphere, between 6 and 30 miles up in the atmosphere, and in the ocean depths, down to 32,800 feet or 10,000 meters deep.

Aerobes, or aerobic bacteria, can only grow where there is oxygen. Some types can cause problems for the human environment, such as corrosion, fouling, problems with water clarity, and bad smells.

Anaerobes, or anaerobic bacteria, can only grow where there is no oxygen. In humans, this is mostly in the gastrointestinal tract. They can also cause gas gangrene, tetanus, botulism, and most dental infections.

Bacteria can thrive even in extreme environments, such as glaciers.

Facultative anaerobes, or facultative anaerobic bacteria, can live either with or without oxygen, but they prefer environments where there is oxygen. They are mostly found in soil, water, vegetation and some normal flora of humans and animals. Examples include Salmonella.

Mesophiles, or mesophilic bacteria, are the bacteria responsible for most human infections. They thrive in moderate temperatures, around 37 degrees Celsius. This is the temperature of the human body.

Examples include Listeria monocytogenes, Pesudomonas maltophilia, Thiobacillus novellus, Staphylococcus aureus, Streptococcus pyrogenes, Streptococcus pneumoniae, Escherichia coli, and Clostridium kluyveri.

The human intestinal flora, or gut microbiome, contains beneficial mesophilic bacteria, such as dietary Lactobacillus acidophilus.

Extremophiles, or extremophilic bacteria, can withstand conditions considered too extreme for most life forms.

Thermophiles can live in high temperatures, up to 75 to 80 degrees Celsius (C), and hyperthermophiles can surivive in temperatures up to 113 degrees C.

Deep in the ocean, bacteria live in total darkness by thermal vents, where both temperature and pressure are high. They make their own food by oxidizing sulfur that comes from deep inside the earth.

Other extremophiles include:

- Halophiles, found only in a salty environment

- Acidophiles, some of which live in environments as acidic as pH0

- Alkaliphiles, living in alkiline environments up to pH 10.5

- Psychrophiles, found in cold temperatures, for example, in glaciers

Extremophiles can survive where no other organism can.

Reproduction and Transformation

Bacteria may reproduce and change using the following methods:

- Binary fission: An asexual form of reproduction, in which a cell continues to grow until a new cell wall grows through the center, forming two cells. These separate, making two cells with the same genetic material.

- Transfer of genetic material: Cells acquire new genetic material through processes known as conjugation, transformation, or transduction. These processes can make bacteria stronger and more able to resist threats, such as antibiotic medication.

- Spores: When some types of bacteria are low on resources, they can form spores. Spores hold the organism's DNA material and contain the enzymes needed for germination. They are very resistant to environmental stresses. The spores can remain inactive for centuries, until the right conditions occur. Then they can reactivate and become bacteria.

- Spores can survive through periods of environmental stress, including ultraviolet (UV) and gamma radiation, desiccation, starvation, chemical exposure, and extremes of temperature.

Some bacteria produce endospores, or internal spores, while others produce exospores, which are released outside. These are known as cysts.

Clostridium is an example of an endospore-forming bacterium. There are about 100 species of Clostridium, including Clostridium botulinim (C. botulinim) or botulism, responsible for a potentially fatal kind of food poisoning, and Clostridium difficile (C. Difficile), which causes colitis and other intestinal problems.

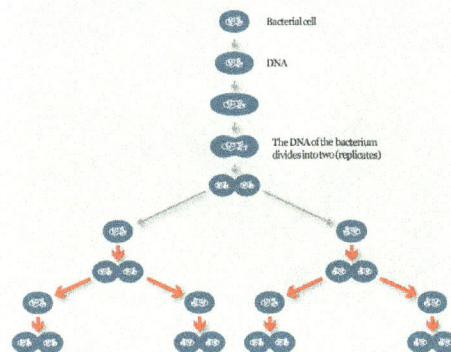

Reproduction through Binary Fission
One becomes two ... two becomes four ... four becomes eight
This means in 8 hours, the number of bacteria will have risen to a colossal 16,777216.

Uses

Bacteria are often thought of as bad, but many are helpful. We would not exist without them. The oxygen we breathe was probably created by the activity of bacteria.

Human Survival

Many of the bacteria in the body play an important role in human survival. Bacteria in the digestive system break down nutrients, such as complex sugars, into forms the body can use.

Non-hazardous bacteria also help prevent diseases by occupying places that the pathogenic, or disease-causing, bacteria want to attach to. Some bacteria protect us from disease by attacking the pathogens.

Nitrogen Fixation

Bacteria take in nitrogen and release it for plant use when they die. Plants need nitrogen in the soil to live, but they cannot do this themselves. To ensure this, many plant seeds have a small container of bacteria that is used when the plant sprouts.

Food Technology

Lactic acid bacteria, such as Lactobacillus and Lactococcus together with yeast and molds, or fungi, are used to prepare foods such as as cheese, soy sauce, natto (fermented soy beans), vinegar, yoghurt, and pickles.

Not only is fermentation useful for preserving foods, but some of these foods may offer health benefits.

For example, some fermented foods contain types of bacteria that are similar to those linked with gastrointestinal health. Some fermentation processes lead to new compounds, such as lactic acid, which that appear to have an anti-inflammatory effect.

Cheese making involves bacteria.

Bacteria in Industry and Research

Bacteria can break down organic compounds. This is useful for activities such as waste processing and cleaning up oil spills and toxic waste.

The pharmaceutical and chemical industries use bacteria in the production of certain chemicals.

Bacteria are used in molecular biology, biochemistry and genetic research, because they can grow quickly and are relatively easy to manipulate. Scientists use bacteria to study how genes and enzymes work.

Bacteria are needed to make antibiotics.

Bacillus thuringiensis (BT) is a bacterium that can be used in agriculture instead of pesticides. It does not have the undesirable environmental consequences associated with pesticide use.

Hazards

Some types of bacteria can cause diseases in humans, such as cholera, diptheria, dysentery, bubonic plague, pneumonia, tuberculosis (TB), typhoid, and many more.

If the human body is exposed to bacteria that the body does not recognize as helpful, the immune system will attack them. This reaction can lead to the symptoms of swelling and inflammation that we see, for example, in an infected wound.

Resistance

In 1900, pneumonia, TB, and diarrhea were the three biggest killers in the United States (U.S.). Sterilization techniques and antibiotic medications have led to a significant drop in deaths from bacterial diseases.

However, the overuse of antibiotics is making bacterial infection harder to treat. As the bacteria mutate, they become more resistant to existing antibiotics, making infections harder to treat. Bacteria transform naturally, but the overuse of antibiotics is speeding up this process.

"Even if new medicines are developed, without behaviour change, antibiotic resistance will remain a major threat." World Health Organization (WHO).

For this reason, scientists and health authorities are calling on doctors not to prescribe antibiotics unless it is necessary, and for people to practice other ways of preventing disease, such as good food hygiene, hand washing, vaccination, and safe sex.

The Gut Microbiome

Recent research has led to a new and growing awaress of how the human body interacts with bacteria, and particularly the communities of bacteria living in the intestinal tract, known as the gut microbiome, or gut flora.

In 2009, researchers published findings suggesting that women with obesity were more likely to have a particular kind of bacteria Selenomonas noxia (S. noxia) in their mouth.

In 2015, scientists at the University of North Carolina found that the intestines of people with anorexia contain "very different" bacteria, or microbial commiunities, compared with people who do not have the condition. They suggest that this may have a psychological impact.

Bacteriology

Bacteriology is the study of bacteria. Bacteria are microscopic organisms composed of a single cell. They are generally referred to as microorganisms because they are so tiny that a microscope is often needed to visualize them. An individual who studies, identifies, and classifies bacteria is called a bacteriologist. He usually does his studies in the laboratory.

The microscope is an essential tool for many bacteriologists as it can magnify the minute organisms many times their actual size. The improvement of the microscope by Anton van Leeuwenhoek has opened the minute world of bacteria to everyone. It was in 1676 when Leeuwenhoek first discovered bacteria.

Different classes of bacteria have different requirements for growth. Some cannot survive extremes of temperatures, while others prefer very low or high temperatures. Many bacteria also differ in their oxygen needs and nutrient needs. Other ways to identify bacteria are through their appearance or shape, the substances they produce, and through their chemical reactions when tested in the laboratory. For example, rod-shaped bacteria are called bacilli, while round-shaped bacteria are known as cocci.

In bacteriology, the structure, functions, and growth of various bacteria have been discovered. Bacteriology has also explored the positive and negative impact of bacteria in the environment and in human beings. Another important function is the identification of bacteria that often cause disease in man and animals, and the mechanisms of how they bring about infection. This is an important aspect of bacteriology, which leads to the development of antibiotics or antibacterial drugs known to treat diseases caused by bacteria.

Bacteriology is a subcategory of microbiology, the study of microorganisms. Aside from bacteria, microbiology also studies fungi, viruses, and parasites in association to the diseases they cause in man. In medicine, microbiology and immunology are often studied together. Immunology deals with the responses of the immune system to the presence of microorganisms inside the body. Treatment and prevention of diseases are made possible because of these studies.

Patients suspected of having infectious diseases are often requested to submit samples such as blood, urine, sputum, and feces, for examination. In the laboratory, bacteriologists then grows the bacteria present in the sample by planting them in certain growth media. Strict and sterile procedures are usually observed in growing the bacteria in order to isolate the bacteria causing the disease and to prevent the bacteria from spreading around the laboratory. Once bacteria are identified, a proper diagnosis can be done and patients can be given the right antibiotic for treatment.

References

- Microbiology: aboutbioscience.org, Retrieved 14 June 2018

- Bacteria, introducing-microbes, about-microbiology: microbiologyonline.org, Retrieved 11 July 2018

- Bacteria-51641: livescience.com, Retrieved 23 April 2018

- What-is-bacteriology: wisegeek.com, Retrieved 16 March 2018

Chapter 2

Cell Structure of Bacteria

The bacterial cell is encapsulated inside a cell membrane that is composed of phospholipids and acts as a barrier to contain proteins, nutrients and other essential components of the cytoplasm. The topics elucidated in this chapter cover some of the important components in the cell structure of bacteria, such as the flagella, plasmid ribosome, mesosomes, gas vacuoles, etc.

Bacteria Cell Structure are as unrelated to human beings as living things can be, but bacteria are essential to human life and life on planet Earth. Although they are notorious for their role in causing human diseases, from tooth decay to the Black Plague, there are beneficial species that are essential to good health.

Prokaryotic Cell Structure

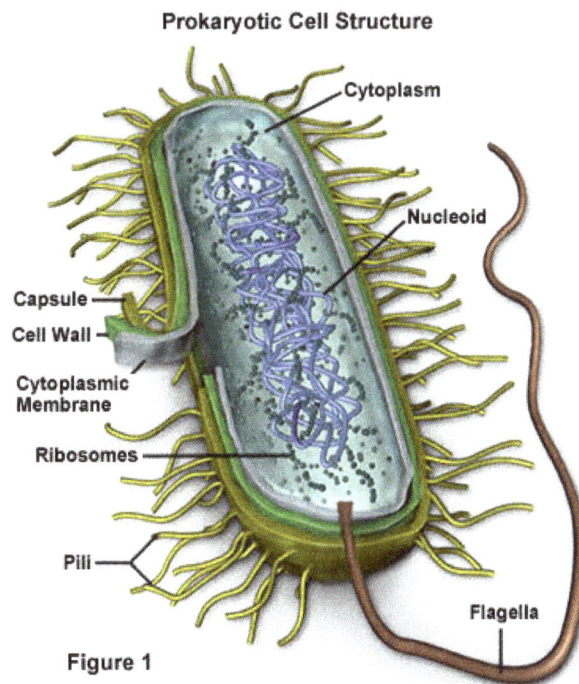

Figure 1

For example, one species that lives symbiotically in the large intestine manufactures vitamin K, an essential blood clotting factor. Other species are beneficial indirectly. Bacteria give yogurt its tangy flavor and sourdough bread its sour taste. They make it possible for ruminant animals (cows, sheep, goats) to digest plant cellulose and for some plants, (soybean, peas, alfalfa) to convert nitrogen to a more usable form.

Bacteria are prokaryotes, lacking well-defined nuclei and membrane-bound organelles, and with chromosomes composed of a single closed DNA circle. They come in many shapes and sizes, from minute spheres, cylinders and spiral threads, to flagellated rods, and filamentous chains. They are found practically everywhere on Earth and live in some of the most unusual and seemingly inhospitable places.

Evidence shows that bacteria were in existence as long as 3.5 billion years ago, making them one of the oldest living organisms on the Earth. Even older than the bacteria are the archeans (also called archaebacteria) tiny prokaryotic organisms that live only in extreme environments: boiling water, super-salty pools, sulfur-spewing volcanic vents, acidic water, and deep in the Antarctic ice. Many scientists now believe that the archaea and bacteria developed separately from a common ancestor nearly four billion years ago. Millions of years later, the ancestors of today's eukaryotes split off from the archaea. Despite the superficial resemblance to bacteria, biochemically and genetically, the archea are as different from bacteria as bacteria are from humans.

In the late 1600s, Antoni van Leeuwenhoek became the first to study bacteria under the microscope. During the nineteenth century, the French scientist Louis Pasteur and the German physician Robert Koch demonstrated the role of bacteria as pathogens (causing disease). The twentieth century saw numerous advances in bacteriology, indicating their diversity, ancient lineage, and general importance. Most notably, a number of scientists around the world made contributions to the field of microbial ecology, showing that bacteria were essential to food webs and for the overall health of the Earth's ecosystems. The discovery that some bacteria produced compounds lethal to other bacteria led to the development of antibiotics, which revolutionized the field of medicine.

There are two different ways of grouping bacteria. They can be divided into three types based on their response to gaseous oxygen. Aerobic bacteria require oxygen for their health and existence and will die without it. Anerobic bacteria can't tolerate gaseous oxygen at all and die when exposed to it. Facultative aneraobes prefer oxygen, but can live without it.

The second way of grouping them is by how they obtain their energy. Bacteria that have to consume and break down complex organic compounds are heterotrophs. This includes species that are found in decaying material as well as those that utilize fermentation or respiration. Bacteria that create their own energy, fueled by light or through chemical reactions, are autotrophs.

Cell Envelope

The bacterial cell surface (or envelope) can vary considerably in its structure, and it plays a central role in the properties and capabilities of the cell. The one feature present in all cells is the cytoplasmic membrane, which separates the inside of the cell from its external environment, regulates the flow of nutrients, maintains the proper intracellular milieu, and prevents the loss of the cell's contents. The cytoplasmic membrane carries out many necessary cellular functions, including energy generation, protein secretion, chromosome segregation, and efficient active transport of nutrients. It is a typical unit membrane composed of proteins and lipids, basically similar to the membrane that surrounds all eukaryotic cells. It appears in electron micrographs as a triple-layered structure of lipids and proteins that completely surround the cytoplasm.

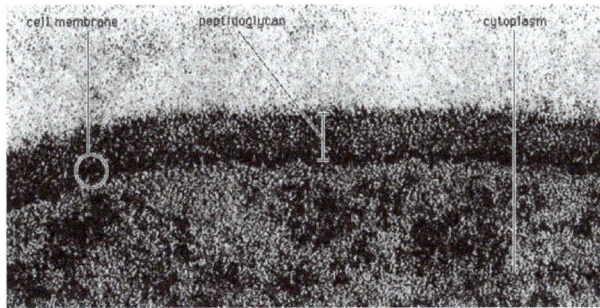

peptidoglycan layer of Bacillus coagulans
A portion of the gram-positive bacterium Bacillus coagulans showing the
cell wall's thick peptidoglycan layer that surrounds the cell membrane.

Lying outside of this membrane is a rigid wall that determines the shape of the bacterial cell. The wall is made of a huge molecule called peptidoglycan (or murein). In gram-positive bacteria the peptidoglycan forms a thick meshlike layer that retains the blue dye of the Gram stain by trapping it in the cell. In contrast, in gram-negative bacteria the peptidoglycan layer is very thin (only one or two molecules deep), and the blue dye is easily washed out of the cell.

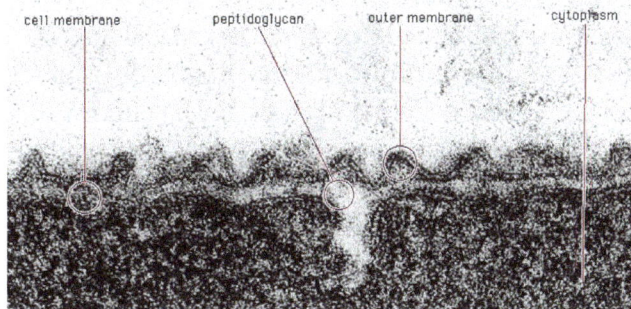

peptidoglycan layer of Aquaspirillum serpens
The gram-negative bacterium Aquaspirillum serpens has a thin peptidoglycan
layer that lies between the cell membrane and the outer membrane.

Peptidoglycan occurs only in the Bacteria (except for those without a cell wall, such as Mycoplasma). Peptidoglycan is a long-chain polymer of two repeating sugars (N-acetylglucosamine and N-acetyl muramic acid), in which adjacent sugar chains are linked to one another by peptide bridges that confer rigid stability. The nature of the peptidebridges differs considerably between species of bacteria but in general consists of four amino acids: L-alanine linked to D-glutamic acid, linked to either diaminopimelic acid in gram-negative bacteria or L-lysine, L-ornithine, or diaminopimelic acid in gram-positive bacteria, which is finally linked to D-alanine. In gram-negative bacteria the peptide bridges connect the D-alanine on one chain to the diaminopimelic acid on another chain. In gram-positive bacteria there can be an additional peptide chain that extends the reach of the cross-link; for example, there is an additional bridge of five glycines in Staphylococcus aureus.

Peptidoglycan synthesis is the target of many useful antimicrobial agents, including the β-lactam antibiotics (e.g., penicillin) that block the cross-linking of the peptide bridges. Some of the proteins that animals synthesize as natural antibacterial defense factors attack the cell walls of bacteria. For example, an enzyme called lysozyme splits the sugar chains that are the backbone of peptidoglycan molecules. The action of any of these agents weakens the cell wall and disrupts the bacterium.

In gram-positive bacteria the cell wall is composed mainly of a thick peptidoglycan meshwork interwoven with other polymers called teichoic acids (from the Greek word teichos, meaning "wall") and some proteins or lipids. In contrast, gram-negative bacteria have a complex cell wall that is composed of multiple layers in which an outer membrane layer lies on top of a thin peptidoglycan layer. This outer membrane is composed of phospholipids, which are complex lipids that contain molecules of phosphate, and lipopolysaccharides, which are complex lipids that are anchored in the outer membrane of cells by their lipid end and have a long chain of sugars extending away from the cell into the medium. Lipopolysaccharides, often called endotoxins, are toxic to animals and humans; their presence in the bloodstream can cause fever, shock, and even death. For most gram-negative bacteria, the outer membrane forms a barrier to the passage of many chemicals that would be harmful to the bacterium, such as dyes and detergentsthat normally dissolve cellular membranes. Impermeability to oil-soluble compounds is not seen in other biological membranes and results from the presence of lipopolysaccharides in the membrane and from the unusual character of the outer membrane proteins. As evidence of the ability of the outer membrane to confer resistance to harsh environmental conditions, some gram-negative bacteria grow well in oil slicks, jet fuel tanks, acid mine drainage, and even bottles of disinfectants.

The Archaea have markedly different surface structures from the Bacteria. They do not have peptidoglycan; instead, their membrane lipids are made up of branched isoprenoidslinked to glycerol by ether bonds. Some archaea have a wall material that is similar to peptidoglycan, except that the specific sugar linked to the amino acid bridges is not muramic acid but talosaminuronic acid. Many other archaeal species use proteins as the basic constituent of their walls, and some lack a rigid wall.

Bacterial Capsule

Most procaryotes contain some sort of a polysaccharide layer outside of the cell wall polymer. In a general sense, this layer is called a capsule. A true capsule is a discrete detectable layer of polysaccharides deposited outside the cell wall. A less discrete structure or matrix which embeds the cells is a called a slime layer or a biofilm. A type of capsule found in bacteria called a glycocalyx is a thin layer of tangled polysaccharide fibers which occurs on surface of cells growing in nature (as opposed to the laboratory). Some microbiologists refer to all capsules as glycocalyx and do not differentiate microcapsules.

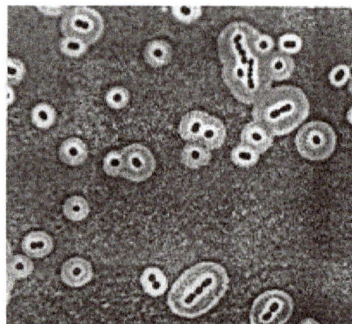

Figure: Bacterial capsules outlined by India ink viewed by light microscopy. This is a true capsule, a discrete layer of polysaccharide surrounding the cells. Sometimes bacterial cells are embedded more randomly in a polysaccharide matrix called a slime layer or biofilm. Polysaccharide films that may inevitably be present on the surfaces of bacterial cells, but which cannot be detected visually, are called glycocalyx.

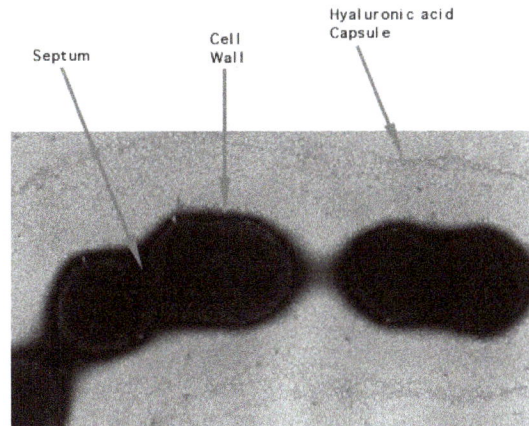

Figure: Negative stain of *Streptococcus pyogenes* viewed by transmission electron microscopy (28,000X). The halo around the chain of cells is the hyaluronic acid capsule that surrounds the exterior of the bacteria. The septa between dividing pairs of cells may also be seen.

Capsules are generally composed of polysaccharide; rarely they contain amino sugars or peptides.

Table: Chemical composition of some bacterial capsules.

Bacterium	Capsule composition	Structural subunits
Gram-positive Bacteria		
Bacillus anthracis	polypeptide (polyglutamic acid)	D-glutamic acid
Bacillus megaterium	polypeptide and polysaccharide	D-glutamic acid, amino sugars, sugars
Streptococcus mutans	polysaccharide	(dextran) glucose
Streptococcus pneumoniae	polysaccharides	sugars, amino sugars, uronic acids
Streptococcus pyogenes	polysaccharide (hyaluronic acid)	N-acetyl-glucosamine and glucuronic acid
Gram-negative Bacteria		
Acetobacter xylinum	polysaccharide	(cellulose) glucose
Escherichia coli	polysaccharide (colonic acid)	glucose, galactose, fucose glucuronic acid
Pseudomonas aeruginosa	polysaccharide	mannuronic acid
Azotobacter vinelandii	polysaccharide	glucuronic acid
Agrobacterium tumefaciens	polysaccharide	(glucan) glucose

Capsules have several functions and often have multiple functions in a particular organism. Like fimbriae, capsules, slime layers, and glycocalyx often mediate adherence of cells to surfaces. Capsules also protect bacterial cells from engulfment by predatory protozoa or white blood cells (phagocytes), or from attack by antimicrobial agents of plant or animal origin. Capsules in certain soil bacteria protect cells from perennial effects of drying or desiccation. Capsular materials (e.g. dextrans) may be overproduced when bacteria are fed sugars to become reserves of carbohydrate for subsequent metabolism.

Figure: Colonies of Bacillus anthracis. The slimy or mucoid appearance of a bacterial colony is usually evidence of capsule production. In the case of B. anthracis, the capsule is composed of poly-D-glutamate. The capsule is an essential determinant of virulence to the bacterium. In the early stages of colonization and infection the capsule protects the bacteria from assaults by the immune and phagocytic systems.

Some bacteria produce slime materials to adhere and float themselves as colonial masses in their environments. Other bacteria produce slime materials to attach themselves to a surface or substrate. Bacteria may attach to surface, produce slime, divide and produce microcolonies within the slime layer, and construct a biofilm, which becomes an enriched and protected environment for themselves and other bacteria.

Importance of Bacterial Capsule

1. Virulence determinants: Capsules are anti-phagocytic. They limit the ability of phagocytes to engulf the bacteria. The smooth nature and negative charge of the capsule prevents the phagocyte from adhering to and engulfing the bacterial cell. If a pathogenic bacteria lose capsule (by mutation), they wont be able to cause disease (i.e. loses disease causing capacity).

2. Saving engulfed bacteria from the action of neutrophil: Bacterial capsule prevents the direct access of lysosome contents with the bacterial cell, preventing their killing.

3. Prevention of complement-mediated bacterial cell lysis.

4. Protection of anaerobes from oxygen toxicity.

5. Identification of bacteria:

   i. Using specific antiserum against capsular polysaccharide. E.g. Quellung reaction.

   ii. Colony characteristics in culture media: Bacteria with capsules form smooth (S) colonies while those without capsules form rough (R) colonies. A given bacterial species may undergo a phenomenon called S-R variation whereby the cell loses the ability to form a capsule. Some capsules are very large and absorb water; bacteria with this type of capsule (e.g., Klebsiella pneumoniae) form mucoid (M) colonies.

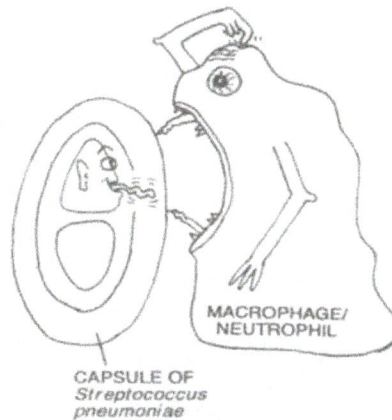

CAPSULE OF
Streptococcus
pneumoniae

MACROPHAGE/
NEUTROPHIL

Anti-phagocytic nature of Bacterial capsule

6. Development of Vaccines: Capsular polysaccharides are used as the antigens in certain vaccines. For examples:

    • Polyvalent (23 serotypes) polysaccharide vaccine of Streptococcus pneumoniae capsule.

    • Polyvalent (4 serotypes) vaccine of Neisseria meningitidis capsule.

    • A monovalent vaccine made up of capsular material from Haemophilus influenzae.

7. Initiation of infection: Capsules helps the organism to adhere to host cells.The capsule also facilitates and maintains bacterial colonization of biologic (e.g. teeth) and inanimate (e.g. prosthetic heart valves) surfaces through formation of biofilms.

8. Receptors for Bacteriophages.

Cell Wall

Bacteria are protected by a rigid cell wall composed of peptidoglycans.

Bacterial cells lack a membrane bound nucleus. Their genetic material is naked within the cytoplasm. Ribosomes are their only type of organelle. The term "nucleoid" refers to the region of the cytoplasm where chromosomal DNA is located, usually a singular, circular chromosome. Bacteria are usually single-celled, except when they exist in colonies. These ancestral cells reproduce by means of binary fission, duplicating their genetic material and then essentially splitting to form two daughter cells identical to the parent. A wall located outside the cell membrane provides the cell support, and protection against mechanical stress or damage from osmotic rupture and lysis. The major component of the bacterial cell wall is peptidoglycan or murein. This rigid structure of peptidoglycan, specific only to prokaryotes, gives the cell shape and surrounds the cytoplasmic membrane. Peptidoglycan is a huge polymer of disaccharides (glycan) cross-linked by short chains of identical amino acids (peptides) monomers. The backbone of the peptidoglycan molecule is composed of two derivatives of glucose: N-acetylglucosamine (NAG) and N-acetlymuramic acid (NAM) with a pentapeptide coming off NAM and varying slightly among bacteria. The NAG and

NAM strands are synthesized in the cytosol of the bacteria. They are connected by inter-peptide bridges. They are transported across the cytoplasmic membrane by a carrier molecule called bactoprenol. From the peptidoglycan inwards all bacterial cells are very similar. Going further out, the bacterial world divides into two major classes: Gram positive (Gram +) and Gram negative (Gram -). The cell wall provides important ligands for adherence and receptor sites for viruses or antibiotics.

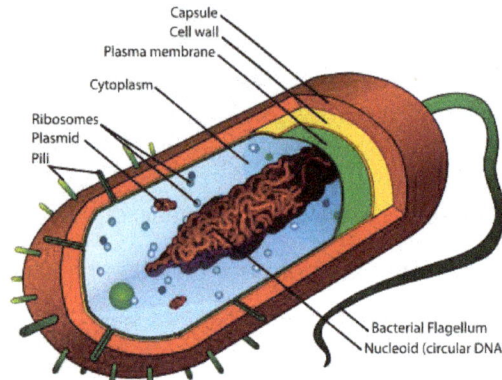

Bacterial Cell Wall: The anatomy of bacterial cell structure.

Gram-negative Outer Membrane

The Gram-negative cell wall is composed of an outer membrane, a peptidoglygan layer, and a periplasm.

In the Gram-negative Bacteria the cell wall is composed of a single layer of peptidoglycan surrounded by a membranous structure called the outer membrane. The gram-negative bacteria do not retain crystal violet but are able to retain a counterstain, commonly safranin, which is added after the crystal violet. The safranin is responsible for the red or pink color seen with a gram-negative bacteria. The Gram-negative's cell wall is thinner (10 nanometers thick) and less compact than that of Gram-positive bacteria, but remains strong, tough, and elastic to give them shape and protect them against extreme environmental conditions. The outer membrane of Gram-negative bacteria invariably contains a unique component, lipopolysaccharide (LPS) in addition to proteins and phospholipids. The LPS molecule is toxic and is classified as an endotoxin that elicits a strong immune response when the bacteria infect animals.

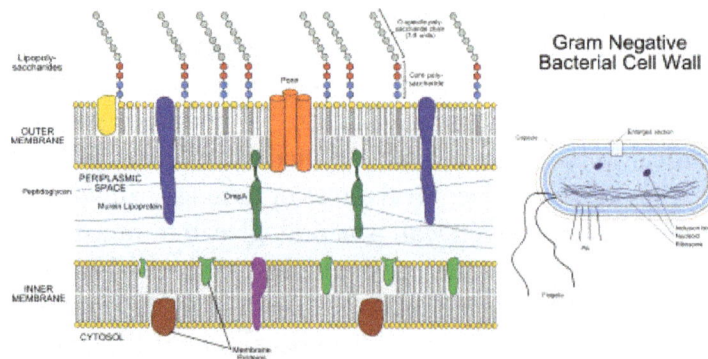

Structure of Gram-negative cell wall: Gram-negative outer membrane composed of lipopolysaccharides.

In Gram-negative bacteria the outer membrane is usually thought of as part of the outer leaflet of the membrane structure and is relatively permeable. It contains structures that help bacteria adhere to animal cells and cause disease. The peptidoglycan layer is non-covalently anchored to lipoprotein molecules called Braun's lipoproteins through their hydrophobic head. Sandwiched between the outer membrane and the plasma membrane, a concentrated gel-like matrix (the periplasm) is found in the periplasmic space. It is in fact an integral compartment of the gram-negative cell wall and contains binding proteins for amino acids, sugars, vitamins, iron, and enzymes essential for bacterial nutrition. The periplasm space can act as reservoir for virulence factors and a dynamic flux of macromolecules representing the cell's metabolic status and its response to environmental factors. Together, the plasma membrane and the cell wall (outer membrane, peptidoglycan layer, and periplasm) constitute the gram-negative envelope.

Gram-positive Cell Envelope

Gram-positive bacteria have cell envelopes made of a thick layer of peptidoglycans.

Gram-positive bacteria are stained dark blue or violet by Gram staining. While Gram staining is a valuable diagnostic tool in both clinical and research settings, not all bacteria can be definitively classified by this technique, thus forming Gram-variable and Gram-indeterminate groups as well.

It is based on the chemical and physical properties of their cell walls. Primarily, it detects peptidoglycan, which is present in a thick layer in Gram-positive bacteria. A Gram-positive results in a purple/blue color while a Gram-negative results in a pink/red color. The Gram stain is almost always the first step in the identification of a bacterial organism, and is the default stain performed by laboratories over a sample when no specific culture is referred.

In Gram-positive bacteria, the cell wall is thick (15-80 nanometers), and consists of several layers of peptidoglycan. They lack the outer membrane envelope found in Gram-negative bacteria. Running perpendicular to the peptidoglycan sheets is a group of molecules called teichoic acids, which are unique to the Gram-positive cell wall. Teichoic acids are linear polymers of polyglycerol or polyribitol substituted with phosphates and a few amino acids and sugars.

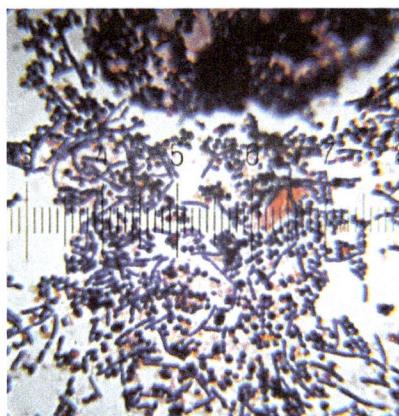

Gram-positive bacteria: These bacteria stain violet by Gram staining.

The teichoic acid polymers are occasionally anchored to the plasma membrane (called lipoteichoic acid, LTA), and apparently directed outward at right angles to the layers of peptidoglycan. Teichoic acids give the Gram-positive cell wall an overall negative charge due to the presence of phospho-diester bonds between teichoic acid monomers. The functions of teichoic acid are not fully known but it is believed to serve as a chelating agent and means of adherence for the bacteria. These are essential to the viability of Gram-positive bacteria in the environment and provide chemical and physical protection.

One idea is that they provide a channel of regularly-oriented, negative charges for threading positively-charged substances through the complicated peptidoglycan network. Another theory is that teichoic acids are in some way involved in the regulation and assembly of muramic acid sub-units on the outside of the plasma membrane.

There are instances, particularly in the streptococci, wherein teichoic acids have been implicated in the adherence of the bacteria to tissue surfaces and are thought to contribute to the pathogenicity of Gram-positive bacteria.

Mycoplasmas and Other Cell-wall-deficient Bacteria

Some bacteria lack a cell wall but retain their ability to survive by living inside another host cell.

For most bacterial cells, the cell wall is critical to cell survival, yet there are some bacteria that do not have cell walls. Mycoplasmaspecies are widespread examples and some can be intracellular pathogens that grow inside their hosts. This bacterial lifestyle is called parasitic or saprophytic. Cell walls are unnecessary here because the cells only live in the controlled osmotic environment of other cells. It is likely they had the ability to form a cell wall at some point in the past, but as their lifestyle became one of existence inside other cells, they lost the ability to form walls.

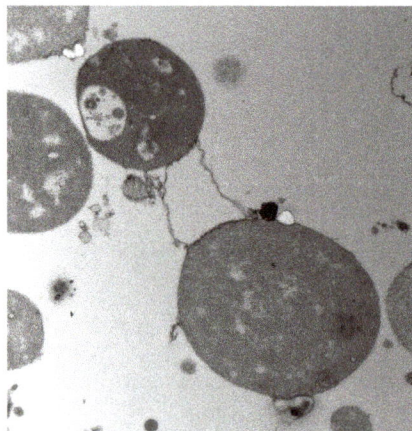

L-form bacteria: L-form bacterial lack a cell wall structure.

Consistent with this very limited lifestyle within other cells, these microbes also have very small genomes. They have no need for the genes for all sorts of biosynthetic enzymes, as they can steal the final components of these pathways from the host. Similarly, they have no need for genes encoding many different pathways for various carbon, nitrogen and energy sources, since their intracellular environment is completely predictable. Because of the absence of cell walls, Mycoplasma have a spherical shape and are quickly killed if placed in an environment with very high or very low salt concentrations. However, Mycoplasmado have unusually tough membranes that are more resistant to rupture than other bacteria since this cellular membrane has to contend with the host cell factors. The presence of sterols in the membrane contributes to their durability by helping to increase the forces that hold the membrane together. Other bacterial species occasionally mutate or respond to extreme nutritional conditions by forming cells lacking walls, termed L-forms. This phenomenon is observed in both gram-positive and gram-negative species. L-forms have varied shapes and are sensitive to osmotic shock.

Cell Walls of Archaea

Archaeal cell walls differ from bacterial cell walls in their chemical composition and lack of peptidoglycans.

As with other living organisms, archaeal cells have an outer cell membrane that serves as a protective barrier between the cell and its environment. Within the membrane is the cytoplasm, where the living functions of the archeon take place and where the DNA is located. Around the outside of nearly all archaeal cells is a cell wall, a semi-rigid layer that helps the cell maintain its shape and chemical equilibrium. All three of these regions may be distinguished in the cells of bacteria and most other living organisms.

A closer look at each region reveals structural similarities but major differences in chemical composition between bacterial and archaeal cell wall. Archaea builds the same structures as other organisms, but they build them from different chemical components. For instance, the cell walls of all bacteria contain the chemical peptidoglycan. Archaeal cell walls do not contain this compound, though some species contain a similar one. It is assembled from surface-layer proteins called S-layers. Likewise, archaea do not produce walls of cellulose (as do plants) or chitin (as do fungi). The cell wall of archaeans is chemically distinct. Methanogens are the only exception and possess pseudopeptidoglycan chains in their cell wall that lacks amino acids and N-acetylmuramic acid in their chemical composition. The most striking chemical differences between Archaea and other living things lie in their cell membrane. There are four fundamental differences between the archaeal membrane and those of all other cells: (1) chirality of glycerol, (2) ether linkage, (3) isoprenoid chains, and (4) branching of side chains.

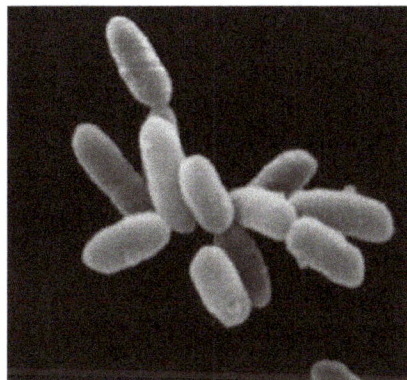

Archaea: Cluster of halobacterium (archaea)

Damage to the Cell Wall

The cell wall is responsible for bacterial cell survival and protection against environmental factors and antimicrobial stress.

The cell wall is the principal stress-bearing and shape-maintaining element in bacteria. Its integrity is thus of critical importance to the viability of a particular cell. In both gram-positive and gram-negative bacteria, the scaffold of the cell wall consists of a cross-linked polymer peptidoglycan. The cell wall of gram-negative bacteria is thin (approximately only 10 nanometers in thickness), and is typically comprised of only two to five layers of peptidoglycan, depending on the growth stage. In gram-positive bacteria, the cell wall is much thicker (20 to 40 nanometers thick).

While the peptidoglycan provides the structural framework of the cell wall, teichoic acids, which make up roughly 50% of the cell wall material, are thought to control the overall surface charge of the wall. This affects murein hydrolase activity, resistance to antibacterial peptides, and adherence to surfaces. Although both of these molecules are polymerized on the surface of the cytoplasmic membrane, their precursors are assembled in the cytoplasm. Any event that interferes with the assembling of the peptidoglycan precursor, and the transport of that object across the cell membrane, where it will integrate into the cell wall, would compromise the integrity of the wall. Damage to the cell wall disturbs the state of cell electrolytes, which can activate death pathways (apoptosis or programmed cell death). Regulated cell death and lysis in bacteria plays an important role in certain developmental processes, such as competence and biofilm development. They also play an important role in the elimination of damaged cells, such as those irreversibly injured by environmental or antibiotic stress. An example of an antibiotic that interferes with bacterial cell wall synthesis is Penicillin. Penicillin acts by binding to transpeptidases and inhibiting the cross-linking of peptidoglycan subunits. A bacterial cell with a damaged cell wall cannot undergo binary fission and is thus certain to die.

Penicillin mechanism of action: Penicillin acts by binding to penicillin binding proteins and inhibiting the cross-linking of peptidoglycan subunits.

Cytoplasm

The cytoplasm is where the organelles carry out the processes necessary for the life of the bacterium.

The six main components of cytoplasm of bacteria. The components are: 1. Ribosomes 2. Molecular Chaperones 3. Nucleoids 4. Plasmids 5. Cytoplasmic Inclusions 6. Spore and Cysts.

Ribosomes

All living cells contain ribosomes which act as a site of protein synthesis. High number of ribosomes represents high rate of protein synthesis and vice versa. Cytoplasm of a prokaryotic cell contains about 10,000 ribosomes which account upto 30% of total dry weight of the cell. Presence of ribosomes in high number gives the cytoplasm a granular appearance.

The eukaryotic ribosomes are found attached to cell membrane, whereas the prokaryotic mesosomes are free in cytoplasm. Prokaryotic ribosomes are smaller and less dense than eukaryotic ribosomes.

Ribosomes of prokaryotes are often called 70S ribosomes and that of eukaryotes as SOS ribosomes. The letter 'S' refers to Svedberg unit which indicates the relative rate of sedimentation during ultracentrifugation. Sedimentation rate depends on size, shape and weight of particles.

Subunits

In general the ultrastructure of ribosomes reveals that these are composed of two subunits, a larger 508 subunit and a smaller 30S subunit. Each subunit is composed of protein and ribosomal RNA (rRNA).

Their association and dissociation depend on the concentration of Mg++ ions. The structure of a ribosome is very complex. The proteins and RNA are inter-wined. James A. Lake presented the structure and function of ribosome.

According to him the smaller subunit of ribosome consists of a head, a base and a platform. With the help of a cleft the platform and head are separated from the base. The larger subunit comprises of a ridge, a central protuberance and a stalk; the former two are separated by a valley.

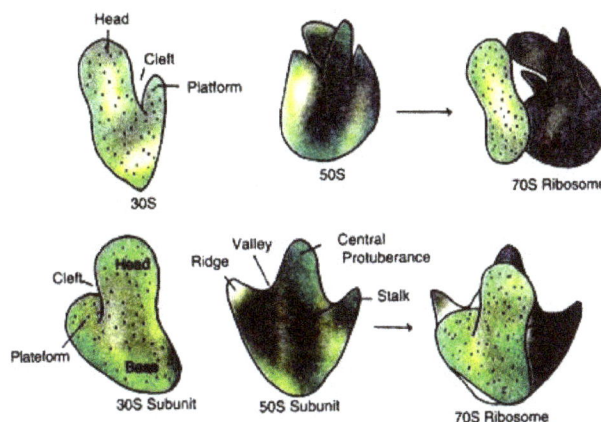

Figure: Three dimensional model of E.coli ribosome shown in two different orientation (A and B)

Chemical Composition

The ribosomes of E. coli consist of three types of RNAs, 5S, 16S and 23S, and 53 proteins. The SOS subunit consists of 5S and 23S RNA, and 34 proteins; 30S subunit consists of 16S RNA and 21 proteins. The 5S RNA is 120 nucleotides long, 16S RNA is about 1,600 nucleotides long and 23S RNA is about 3,200 nucleotides long.

Base sequence of 5S RNA has been strongly conserved throughout the evolution and that of 16S form double stranded hairpin loops. Only 30-35% bases of 16S form single stranded loops. Interactions between rRNA and cellular RNA (mRNA and tRNA) occur.

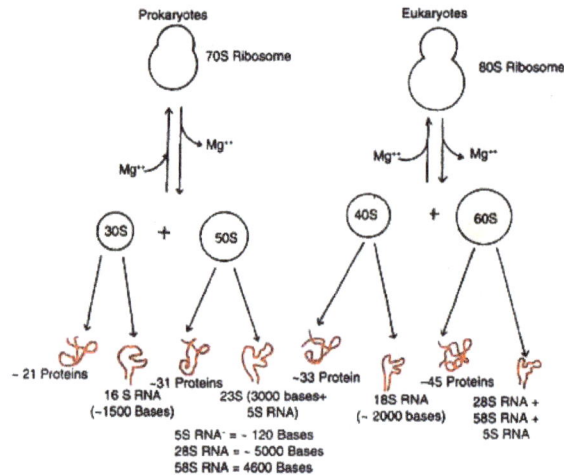

Figure: the rRNA and proteins of prokaryotic (A) and eukaryotic (B) ribosomes.

From 30S RNA of E. coli, 21 proteins have been isolated that are designated as SI to S21. Similarly from larger subunit (50S), 31 proteins (LI to L34) have been isolated. Protein map of ribosome showing their sites on two subunits is presented in figure below.

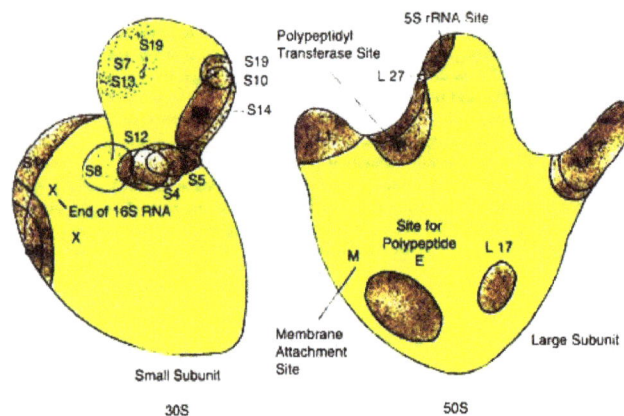

Figure: protein map showing their sites on small subunit and large subunit of ribosomes.

The function of ribosome in protein synthesis is a well established fact. However, they are not specific in nature. Ribosome of one species can be used for protein synthesis in other species.

There are several antibiotics such as streptomycin, neomycin and tetracycline that inhibit protein synthesis on the ribosomes. Even antibodies can kill the prokaryotic microorganisms but not the eukaryotic microorganisms. This is due to differences in prokaryotic and eukaryotic ribosomes.

Molecular Chaperones

It was thought for many years that polypeptides after synthesis fold into native stage and this folding is not determined by its amino acids. It is now clear that there are certain helper proteins called molecular chaperones or chaperones which recognise the newly formed polypeptides and fold to its proper shape.

Proteins fold rapidly into the secondary structure. This unusually open and flexible conformation is called as molten globule. It is the starting points for slow process which results in correct tertiary structure.

There are several chaperones involved in proper protein folding in bacteria. Chaperones were first identified in E. coli mutant that did not allow to replicate, phage lambda. In E. coli at least four chaperones viz., DnaK, DnaJ, GroEL and GroES, and stress protein GrpE are involved in folding process.

They play an important role because after protein synthesis the cytoplasmic matrix is filled with nascent polypeptides and proteins. It is possible that these polypeptides become folded and form a nonfunctional complex. The chaperones check wrong in-folding and promote correct folding. The chaperones are found both in the cells of prokaryotes and eukaryotes.

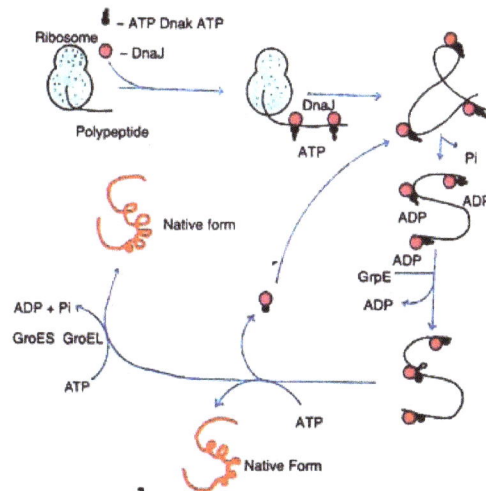

Figure: Mechanism of polypeptide folding by chaperones.

After synthesis of a sufficient length of polypeptide from the ribosome, DnaJ binds to unfolded chain. DnaK complexed with ATP attaches to polypeptide. These two chaperones check the polypeptide from folding. After binding of DnaK to polypeptide ATP is hydrolysed to ADP which increases the ability of DnaK to bind with unfolded peptide.

When polypeptide synthesis is completed, GrpE protein binds to DnaK-polypeptide complex and causes the DnaK to release ADP. Thereafter, ATP binds to DnaK, and DnaK and DnaJ are released from the polypeptide. During these events, polypeptide is folded and reaches to its final native conformation. At this stage if polypeptide is partially folded it binds to DnaJ and DnaK, and repeats the same process again.

Mostly DnaK and DnaJ transfer the polypeptide to GroEL and GroES where final foldings occurs. GroEL is a long hollow barrel shaped complex of 14 subunits which are stacked in two rings, whereas GroES contains four subunits arranged in one ring and can combine to both ends of GroEL. ATP binds to GroEL and changes the ability of the later for polypeptide binding. GroES binds to GroEL and helps in binding and release of refolding peptide.

Heat Shock Proteins

When E. coli cells are exposed to high temperature, metabolic poisons and other stressful conditions, the concentrations of chaperones increase. In E. coli cultures, at temperature between 30 and 40°C, 20 different chaperones often called heat shock proteins are produced within 5 minutes.

These protect the cell from thermal damage and stress, and promote the proper folding of polypeptides. In hypo-thermophiles (e.g. Pyrodictum occultum) that grow at about 110°C, a large amount of chaperones are present.

Mutant E. coli resistant to phage X produces slightly two changed chaperones like heat shock proteins 60 and 70 (hsp60 and hsp70). The eukaryotic cells have families of hsp60 and hsp70 proteins, and different family members function in different organelles.

The mitochondria contain their own hsp60 and hsp70 molecules which are different from those functioning in the cytosol. A special hsp70 helps to fold proteins in the endoplasmic reticulum. The other function of chaperones is the transport of proteins across the membrane.

Nucleoids (Bacterial Chromosome)

As in eukaryotes, in prokaryotes too the basic dye stains the nuclear material and reveals as dense and centrally located bodies of irregular outline. Upon observation with electron microscope it was found that this central region is not separated from the cytoplasm by a membrane and consists of nuclear structure besides the DNA fibrils. The eukaryotes contain a well organized nucleus in which the genetic material is enclosed by a nuclear membrane.

Therefore, the DNA material is not enclosed by any covering. Hence the bacterial chromosome is known as chromatin bodies or nucleoids. The nucleoid is a single long circular double stranded DNA molecule devoid of highly conserved histone protein. The histone is present in eukaryotes, therefore, results the eukaryotic DNA into the beaded structures i.e. nucleosomes.

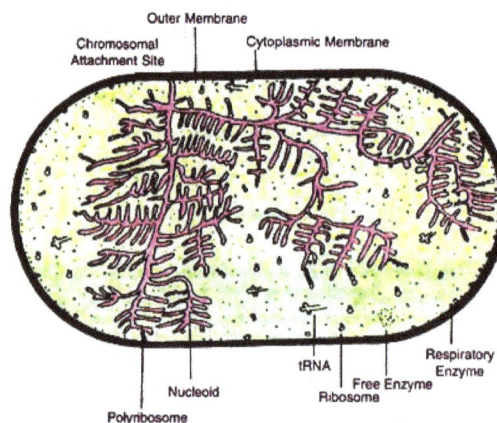

Figure: A cell of E. coil showing nucleoid in centre and the other cell components (diagrammatic).

By the turn of 1960s, the nuclear material was studied in detail. In 1963, J. Cairns succeeded in extracting E. coli DNA under conditions that minimize its shearing. Auto-radiographic studies showed that the DNA was extremely long threads measuring about 1mm in length.

A few threads were circular. Hobot found that the fully extended E. coli DNA was 1 mm long ($4 \times 10^6 \times 3.4$ Å) having molecular weight 3×10^9 Dalton. Through electron microscope they also observed the compaction of chromosome into irregular shaped nucleoids.

The nucleoid is observed as a coralline (coral like) shaped structure the branches of which spread far into the cytoplasm and over the entire area of the cell. By using serial sections it has been possible to reconstruct the ribosome free area of the nucleoid.

Two types of nuclear bodies can be observed, an envelope associated nucleoid and an envelope free nucleoid. Associated with the first type a large amount of RNA, proteins, lipids and peptidoglycan are found, whereas the second type contains less amount of it.

Generally, the number of nucleoid per bacterial cell is one, but in some bacteria the number may go even to four or more. The DNA molecule appears to be present in 10 to 80 super coils. Worcel and Burg proposed the structure of the folded chromosome of E. coli and showed as seven loops, each twisting into a super helix. These loops are held together by a core of DNA.

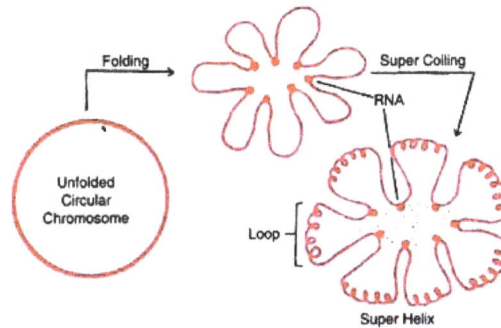

Figure: A model representing the process of folding
and super coiling of bacterial chromosome.

Super coiling may be induced enzymatically. Possibly it may be a factor for the formation of nucleoids. The folded structure was found to be attached to a fragment of cell membrane. This shows that the bacterial chromosome remains associated to a point of cell membrane. This helps in separation of newly replicated DNA molecules.

Plasmids

During 1950s, working on conjugation process it was found that maleness in bacteria is determined by a transmissible genetic element. When male and female bacteria conjugate, every female is converted into a male. This inherited property of male is called the F (fertility) factor which is transmitted by cell to cell contact. Therefore, F is a separate genetic element.

In 1952, J. Lederberg coined the term plasmid as a genetic name for this element. Hence the plasmids may be defined as a small circular, self replicating and double stranded DNA molecule present in bacterial cell, in addition to its chromosome. It replicates independently during cell division and inherited by both of daughter cells. Therefore, its function is not governed by the bacterial chromosome.

In 1960, Jacob Schaeffer and Wollman for the first time used the term episome to denote the extra-chromosomal genetic element that integrated the bacterial chromosome during replication. The number of plasmids ranges from one to hundreds or more per bacterial cell.

A plasmid contains 5-100 genes that determine several biological functions. Under certain circumstances they provide special characteristics to the bacterial cell and help them in survivability. They may even lose without harming the bacterial cell.

Plasmids are the circular DNA molecule but in resting stage helix twists in right hand direction at every 400-600 base pairs and forms supercoils. The twisted form is called covalently closed circular- DNA. After cleaving the twists this form is converted into an open circular form of double stranded DNA molecule.

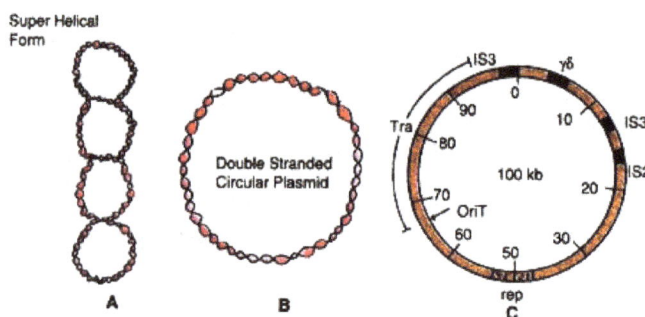

Figure: Plasmid. A, Super helical form; B, double stranded circular form; C, F-plasmid.

Cytoplasmic Inclusions

The cytoplasm of prokaryotic and eukaryotic cells contains several reserve deposits which are called inclusions. Some inclusions are common to most of bacteria and some are restricted to certain species only. These inclusions serve as the basis for identification of bacteria. Shiveley has given an excellent account of inclusion bodies of prokaryotes.

The inclusion bodies are of two types (a) free inclusion bodies (e.g. polyphosphate granules and cyanophycean granules), and (b) single-layered non-unit membrane enclosed inclusion bodies (such as poly P-hydroxybutyrate granules, glycogen granules, sulphur granules, carboxysomes and gas vaculoes). The membrane of inclusion bodies is made up of proteins or lipids.

Bacterial Outer Membrane

The bacterial outer membrane is found in gram-negative bacteria. The outer membrane of Gram-negative bacteria such as Escherichia coli serves as a protective barrier that controls the influx and efflux of solutes. This allows the bacteria to inhabit several different, and often hostile, environments.

The secretion of proteins through outer membrane vesicles (OMVs) is another mechanism, receiving the attention of microbiologists during the past three decades.

OMVs are small spherical bags of 20–300 nm diameters, shed mostly by gram-negative bacteria in the extracellular environment. They are rich in outer membrane proteins and mostly depleted in cytoplasmic contents. They are not products of cell fragmentation or sloughing off the outer membrane but outcome of the normal turnover of the cells. They are not formed by fragmentation of the cell or accidental detachment of some portions of the cell membrane. They are the outcome of the normal turnover of the cells.

Biogenesis of OMVs

Vesicles are formed by protrusion of the outer membrane and pinching-off from the membrane. It is a process which requires energy though no ATP or other energy source is available at the site of budding. Bulging of the membrane can occur when the membrane is disrupted. However, studies on E. coli revealed that the instability of the membrane is not a prerequisite for vesiculation. It is proposed that the imbalance in the turnover of peptidoglycan leads to accumulation of muramic acid. The hydrostatic pressure generated by osmotic imbalance at places in the periplasm leads to the bulging of the outer membrane. Occurrence of low molecular weight muramic acid in some OMVs corroborates this hypothesis.

The bulging of the outer membrane also requires delinking of the outer membrane from peptidoglycan. The proteins linking the outer membrane and peptidoglycan may be disrupted or relocated in the process. This idea was supported by the absence of some outer membrane proteins in OMVs produced by E. coli. It is known that OMVs are produced during normal growth. But their formation has been found to be stimulated in the presence of some physical (heat) and chemical (detergents) agents. Although OMVs are produced within the log phase, some OMVs are also formed during the stationary phase, probably due to the fact that some cells in the bacterial population still remain in the log phase. The effect of nutrients on the production of OMVs is also not the same in all bacteria. Vesicle formation was found to be upregulated in low nutrient condition in case of Lysobacter, whereas it was enhanced in Pseudomonas fragi when nutrients were available.

Studies performed from time to time indicate the role of a number of genes in vesiculation. The Tol-Pal system of E. coli consists of a number of envelope proteins. The genes that encode them are organized in two operons. The system is believed to be involved in maintenance of the integrity of the membrane. Mutations in the Tol-Pal genes are known to be associated with increased formation of OMVs. Mutation in another gene nlp I in E. coli was found to be associated with hypervesiculation without having any effect on the integrity of the membrane. Nlp I is a lipoprotein believed to be involved in cell division. But formation of OMVs is a multifactorial, highly complex phenomenon and extensive studies are required to elucidate its mechanism. its exact role in vesiculation is not yet known. The evidences obtained so far clearly indicate that the formation of OMVs is a multifactorial, highly complex phenomenon and extensive studies are required to elucidate its mechanism.

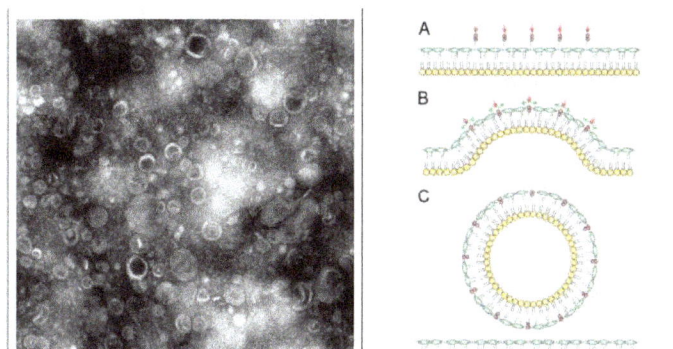

Figure (left): The TEM image of the outermembrane vesicles of S. typhimurium.
Figure (right): Bilayer coupled model of vesicle formation in P. aeruginosa.

Reproduced with permission from J W Schertzer and M Whiteley, A Bilayer-couple model of bacterial outer membrane vesicle biogenesis, mBio.

Structural Components

The outer membrane vesicles of bacteria are chiefly composed of proteins and lipids of the outer membrane and periplasm. The number of proteins differs largely in OMVs obtained from various bacterial strains. In general, in Escherichia coli, 0.2%–0.5% of the outer membrane and periplasm proteins are packaged into OMVs. Studies performed from time to time have revealed the presence of 6 proteins in OMVs obtained from P. aeruginosa, 48 proteins in the OMVs of N. meningitidis, 44 proteins in the OMVs produced by Pseudo altereomonas antarctica NF3 and 141 proteins in the OMVs of E. coli. The porins1 or tarnsmembrane channel proteins are one of the components of OMVs. Among the other proteins, the virulence factors constitute a major fraction. The different types of virulence factors detected in OMVs produced by various pathogenic bacteria include enzymes (protease, acid phosphatase, E-glucuronidase, lipase, urease, cellulase, chitinase), molecular chaperones (Hsp 60), toxins (shiga toxin, heat labile enterotoxin of enterotoxic E. coli, cholera toxin), fimbriae and many other type of proteins.

The composition of the OMVs produced by the same organism also varies with presence and absence of stress. Bacteria under different stress conditions were found to partition more proteins into the vesicles. Growing evidences suggest that cytoplasmic proteins and genetic material like DNA and RNA are also packaged into the vesicles.

Among the lipids, phospholipids and lipopolysaccharides are the major components that are found to occur in OMVs. Occurrence of lipoproteins is also observed in OMVs in some cases. The lipid composition of the OMVs from Pseudomonas syringae Pv tomato T, studied at our laboratory, revealed the presence of phosphatidylethanol amine and phosphatydylglycerol with varying lengths of fatty acyl chains.

Functional Importance

Role in Secretion

OMVs are formed through an energy-requiring process. Hence, their formation and secretion must have some biological significance. It has not been possible to isolate any gram-negative bacterium or its mutant which does not produce OMVs. So vesicle formation might have an indispensible role in survival of the organisms. It appears that bacteria use OMVs as a sort of special secretory device.

There are several features which make them distinct from the other secretory systems. Both soluble and insoluble proteins can be transported to their targets, located far away from the cell, through OMVs. Packaging of proteins into OMVs offers a number of advantages. They remain protected from proteases that might occur in the extracellular environment. Moreover, multiple proteins can be delivered at the target site at a time in required concentration, without being diluted in the extracellular fluid.

A number of models have been proposed so far to explain how OMVs deliver their content into the different types of target cells. It is postulated that OMVs lyse spontaneously to release their content in the vicinity of the target cells. The OMVs are known to be stable and the spontaneous lysis of OMVs is a rare phenomenon. However, the OMVs released by some gram-negative bacteria were found to lyse various gram-positive and gramnegative bacteria by a number of investigators. The vesicles are believed to attach themselves to the cell surface of the grampositive bacteria and break open to release the enzyme peptidoglycan hydrolase, which disrupts the underlying cell-wall. In case of gram-negative bacteria, the vesicles are believed to fuse with the outer membrane of the target cells, release their luminal content in the periplasm, where the enzyme peptidoglycan hydrolase can diffuse around and attack the peptidoglycan in a number of sites.

While interacting with eukaryotic cells (e.g., during pathogen esis), bacterial vesicles are believed to attach themselves to the host cells through some adhesive molecules on their surface and subsequently enter the host cells through endocytosis. For example, in the case of enterotoxigenic Escherichia coli (ETEC) infection, a heat labile enterotoxin (LT), bound to the lipopolysaccharide of the vesicles, gets attached to the GM1 receptor of the host cells. Binding to this receptor is associated with internalization. This is how the vesicles produced by the ETEC appear to enter the intestinal cells. Similarly, a vacuolating toxin produced by Helicobacter pylori is associated with the surface of the OMVs produced by the pathogen. Binding of the vesicles to the host cells and subsequently endocytosis of the vesicles are believed to be mediated through the toxin. However, not all the surface-associated bacterial toxins mediate binding and internalization of the vesicles.

Role in Pathogenesis

OMVs are important for pathogenicity and virulence of bacteria. Studies involving various pathogenic bacteria clearly reveal that they produce OMVs within the infected host tissues. Body fluids (blood, urine) collected from the infected patients in the clinical setup and also from the infected animals in the laboratory are found to contain vesicles, thus indicating migration of OMVs from the site of infection.

Vesicle production in pathogenic strains was found to be several times greater compared to their non-pathogenic counterparts in a number of investigations. Because of their small size, vesicles produced by a pathogenic organism can reach and interact with the host tissues, which are otherwise inaccessible to the bacterium. Some time back, it was demonstrated that the gram-positive bacterium Bacillus anthracis formed OMVs containing toxins and anthrolysin, which were delivered to the host cell. Multivalent adhesins, required for colonization of host tissues by gramnegative bacteria, can be efficiently delivered at the target through OMVs. Both pathogenic and non-pathogenic bacteria produce vesicles for survival inside the host. In non-pathogenic bacteria, they may work by sequestering the attacking bacteriophages.

Role in Biofilm Formation

Biofilms are aggregates of bacteria embedded in a polysaccharide matrix produced by the bacteria. Naturally occurring biofilms observed in various places (e.g., within water pipes or on masts of the ships) harbour multispecies populations. Biofilms formed within the hosts (on urinary catheters, cardiac pacemakers, heart valve replacement, artificial joints and other surgical implants) provide a haven for the pathogenic bacteria; they remain resistant to the antibiotics compared to their free-living counterparts. An abundance of the vesicles is observed in biofilms formed in vivo or in vitro, thus indicating a possible role of OMVs in biofilm formation. They are believed to mediate co-aggregation of bacteria. HmuY, a unique protein produced by Porphyromonas gingivalis, the causative agent of chronic periodontitis, helps the organism to acquire the heme protein and in the formation of the biofilm. This protein was found to be associated with OMVs. The DNA molecules present on the surface of some OMVs are also believed to serve as a bridging component in biofilms. OMVs can also nucleate the process of biofilm formation.

Role in Cell-to-cell Communication

Bacteria communicate among themselves by releasing some soluble chemicals (autoinducers), which are detected by other members of the community, leading to the realization of coordinated behaviour. The amount of autoinducers produced is proportionate to the population size. Thus, autoinducers indicate that the population size has reached a certain threshold (quorum). The phenomenon called quorum sensing is crucially important in biofilm formation, production of virulence factors and antibiotics and also in other physiological processes. While some of the autoinducers are water soluble, a quorum sensing molecule (2- heptyl-3-hydroxy-4-quinolone) produced by Pseudomonas aeruginosa called Pseudomonas Quinolone Signal (PQS) is hydrophobic and, therefore, requires the help of OMVs to be transported from cell-to-cell. Formation of OMVs in this organism on the other hand, is dependent on the synthesis of PQS. Addition of exogenous PQS to a PQS-deficient culture of P. aeruginosa and also to other gram-negative bacteria, leads to the formation of OMVs. Thus it is ensured that the PQS, notwithstanding its hydrophobic nature, reaches its target after formation.

Role in Self Defence

Antibiotics are chemical substances of microbial or synthetic origin. They suppress the growth of or kill microorganisms other than the producer. Hence, antibiotics pose a threat to bacteria. However, bacteria are known to evolve with different types of strategies that make them immune to the growth-inhibitory or killer effects of antibiotics. Recent evidences indicate that the membrane vesicles protect bacteria against some antibiotics. In a study involving a hyper-vesiculating mutant of the wild-type E. coli, two antibiotics acting on the outer membrane (polymixin B and colistin), could not prevent the growth of the mutant. Formation of the vesicles was found to be significantly induced in the presence of the antimicrobial peptides. Growth-inhibitory effect of two membrane-active antibiotics on an Antarctic bacterium was found to be reversed at our laboratory upon addition of OMVs isolated from the same organism, to the culture medium. Uptake of the antibiotics by the vesicles was also demonstrated by an assay involving fluorescent-labelled compounds. Thus, the vesicles appeared to protect the bacterial cells by removing the antibiotics from the medium and not allowing them to reach their target.

On the other hand, presence of ciprofloxacin in the vesicles produced a ciprofloxacin-resistant mutant of the mycoplasma Acholeplasma laidlawii reported some time back, indicates that the vesicles might protect the organism by transporting the antibiotic from inside to the outside of the cells, thus not allowing it to accumulate in sufficient concentration required for its activity. In some cases the OMVs are found to carry the enzyme β -lactamase, which converts β -lactam antibiotics (penicillins and cephalosporins) into therapeutically inactive compounds. The vesicle-transported enzyme is observed to protect not only the producer organism but also some other bacteria known to cooccur with the OMV-producer in the human respiratory tract. The ability of the vesicles to shield the enzyme β -lactamase from inactivation by anti- β -lactamase antibody has been demonstrated in an in vitro study.

In another investigation, two clinical isolates of the bacterium Acinetobacter baumannii, a major cause of healthcare-associated infections, were found to produce OMVs containing the plasmid-borne bla_{OXA-24} gene, which encodes a β -lactamase responsible for carbapenem-resistance. A carbapenem-sensitive strain of the organism was found to turn resistant following incubation with these vesicles. Subsequently, it started releasing OMVs containing the resistance-conferring gene. So in this case, OMVs played the role of a vector for horizontal transfer of the resistanceconferring gene. Thus, OMVs appear to work through multifaceted mechanisms for mediation of antibiotic-resistance in bacteria.

Protective action of OMVs was also demonstrated against the T4 bacteriophage by some other investigators. Besides helping bacteria in containing the challenge posed by the antibiotics and bacteriophages, OMVs also appear to play a role in competition with other bacteria in the extracellular milieu for food. In a nutrient-deficient environment, OMVs produced by P. aeruginosa were found to carry the cell wall degrading enzyme called peptidoglycan hydrolase, which could lyse other bacteria occurring nearby.

Role in Stress Tolerance

Stress is a relative term. Any condition an organism is not normally habituated to endure is a stress for it. Thus, even an aerobic environment is stressful to an obligate anaerobe. However, in a general sense, stress implies extremities of temperature, pressure, pH, salinity, desiccation and radiation. Presence of high concentration of oxidative substances and toxic chemicals, paucity of essential nutrients in the environment and presence of misfolded protein within the cell also exert stress on bacteria.

Proteins, which guide the enzyme RNA polymerase to transcribe the right gene when required, are called sigma factors. It is well known that in response to various types of stress conditions, synthesis of different sigma factors is induced in bacteria. Genes transcribed by them encode proteins, which work in various ways ultimately to relieve the cells of stress. The Sigma E protein is involved in the management of extracytoplasmic stress. It is activated by another protein DegS. In a study conducted at the Duke University Medical Center (Durham, USA) involving the bacterial strain Escherichia coli DH5D, the rate of vesicle formation was found to be enhanced when the temperature was increased. A very high level of vesicle production was observed in a mutant having its degS gene inactivated by transposon mutagenesis. In absence of the DegS protein in this mutant, the stressrelieving effect of the Sigma E was also absent and improper folding of proteins caused by increase in temperature led to cellular stress. Thus, it was evidenced that OMVs also relieve bacterial cells of the stress, caused by the accumulation of misfolded proteins. The investigators also got evidence of a specific packaging mechanism that enabled the misfolded proteins to be incorporated into the OMVs and transported out of the cell. This feature (removal of the undesirable materials from the cell) makes the vesicles analogous to the membrane-bound microparticles of the eukaryotic cells, which serve the same purpose. Overexpression of periplasmic protein also is known to stimulate vesicle production. Increase in vesiculation at high temperature might also be an indirect effect of high temperature, which increases the growth rate leading to enhanced production of OMVs.

Increase in vesiculation is also observed, when bacteria are exposed to various chemical stressors. In a recent investigation, production of OMVs was found to be enhanced in Pseudomonas aeruginosa following treatment with D-cycloserine (a known inhibitor of peptidoglycan synthesis), polymixin B (an antibiotic which acts on the outer membrane of bacteria) and hydrogen peroxide (which imparts oxidative stress inside the bacterial cell). Unlike what was found earlier in E. coli, vesicle formation was not found to increase in P. aeruginosa with increase in temperature. Thus, it is evident that mechanism used for counteracting the same stressor, varies in different gram-negative organisms.

Role in Gene Transfer

OMVs are known to facilitate horizontal gene transfer (HGT, movement of genetic material between different species) by carrying DNA fragments. An example of the ability of the vesicles to facilitate HGT leading to the dissemination of antibiotic resistance in bacteria is highlighted in the relevant portion of this article. However, OMVs need not necessarily contain DNA inside the lumen and internalize DNA fragments into the target cells. The same purpose might be achieved if DNA is carried on the surface of OMVs.

Role in Procurement of Nutrients

Some enzymes (e.g., aminopeptidase) transported by the OMVs help the producer organism in acquiring nutrients from the surrounding. OMVs also carry receptors that acquire nutrients. For example, the pseudomonas quinolone signal (PQS) may act as a siderophore to fetch iron from the surrounding. The PQS-FeOMVs complex delivers iron by being absorbed into the outer membrane by fusion. The complex may also dissociate to release the PQS-bound iron near the cell. Thus, OMVs promote survival of the producer cell in iron-limiting environments, which is most often encountered by pathogenic bacteria inside the host. In an investigation performed a couple of years back, indication was obtained of possible involvement of OMVs in enzymatic reduction and transformation of heavy metals and radionuclides. It was observed that OMVs produced by the metal-reducing bacterium Shewanella contained proteins and cytochromes that are essential for coupling the oxidation of hydrogen to the reduction of multivalent metals; an indication that the OMVs carried some terminal reductase was also obtained.

Biotechnological Potential of OMVs

The ability of OMVs to deliver bacterial toxin in a concentrated form to the host cells makes them a potential candidate for the preparation of vaccines. They are immunogenic, self-adjuvant and easily taken up by the mammalian cells. Hence, several attempts have been made to prepare vaccines against various diseases using OMVs. Cerebrospinal meningitis (most often referred to as meningococcal disease) is a recurrent problem in the developing countries. OMV-vaccines were found to be protective against serogroup B meningococcal outbreak. Subcutaneous injection of OMVs obtained from Burkholderia pseudomallei (the causative agent of melioidosis, a disease responsible for significant morbidity and mortality in Southeast Asia and Northern Australia) was found to be protective against the lethal challenge by B. pseudomallei in aerosol in BALB/c mice.

Pasteurella multocida along with Mannheimia haemolytica is a major cause of bovine respiratory disease. Following intranasal immunization of BALB/c mice using OMVs obtained from these two organisms, humoral and mucosal immune response were evidenced by enzyme-linked immunosorbent assays (ELISA). OMV-immunization was also found to confer immunity against Vibrio cholerae in a rabbit model. These and many other reports in the literature underscore the immense potential of OMVs as a component of vaccines. The vesicle-based vaccines promise safety as they do not contain live bacteria.

In a study reported a couple of years back, production of OMVs was found to be enhanced in Shigella flexneri in presence of gentamicin. The antibiotic-induced vesicles were observed to package the antibiotic, penetrate through S. flexneri-infected cultured human intestinal epithelial cells and kill the intracellular bacteria following delivery of the antibiotic. Gentamicin and other aminoglycoside antibiotics cannot penetrate through mammalian cell membrane and therefore cannot be used to treat infections caused by intracellular pathogens (e.g., Shigella spp, Salmonella spp). This study provides a clue to overcome the problem by using OMVs as a vehicle for drug delivery. In some cases, it has also been possible to transform bacterial cells with DNA packaged into OMVs. So there is multifaceted scope of using OMVs as biotechnological tool. Keeping in mind the involvement of OMVs in pathogenesis and antibiotic-resistance of bacteria, it appears worthwhile to look into the therapeutic potential of the inhibitors of vesicle formation to control bacterial infections.

The various roles of OMVs in bacterial physiology offer an interesting field for investigation. Recent reports on the involvement of OMVs in protection against stress add a new dimension to the present state of knowledge on stress adaptation of bacteria. It has been already mentioned that vesicles are produced also by the archeal and eukaryotic cells. Hence, studies on OMVs are likely to provide insight into one of the basic biological processes that occur in all forms of life. The OMVs are known to interact with both eukaryotic and prokaryotic cells and deliver the contents into the host cells. Taking clues form the identification of different components present in the OMVs, it appears feasible to reconstitute the vesicles with biologically important proteins and to improve their ability to transport the proteins to the host cells.

Fimbriae and Pili

Pili and fibriae are present on the cell surface ad they help in attachment of an organism. They are present in bacteria other then the presence of flagella but they are not used for locomotion. The main difference between pili and fibriae is pili are found only in gram negative bacteria but fibriae is found in both, gram negative and gram positive bacteria.

Pili

Pili is an appendage present on cell surface which is used for the purpose of attachment. They are present only in gram negative bacteria and are long, thick, tubular structure which is made up of protein which is known as pilin. Hence the name, pili. Indirectly, pili are useful in reproduction process of a bacteria because they help in attachment of one bacterium to another during the conjugation process, that is why it is also known as sex-pili. Sex-pili are responsible for the sharing of genes. Plasmid genes are responsible for the manufacture of pili and their number is comparatively less then fimbriae. They are thicker, tube like outgrowths which do not play any role in locomotion. It is also known as organ of adhesion due to its quality of sticking to other cells during the process of conjugation. And while doing it, a pipe like tube called conjugation tube is formed which helps in cell to cell attachment.

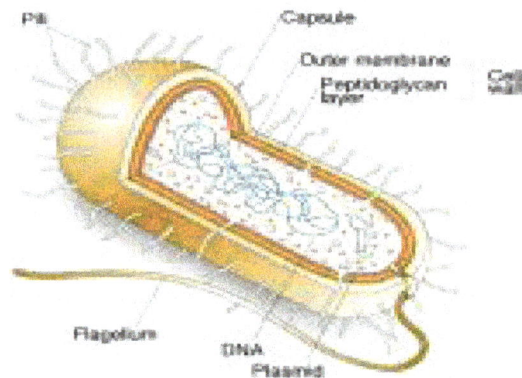

Fimbriae

They are found in gram negative as well as gram positive bacteria but are shorter in length as compared to pili. Bacterial genes which are present in the nucleoid region are basically responsible for the manufacture of fimbriae. Basic function of fimbriae is cell to surface attachment of the bacteria. They are composed of protein sub-units and are evenly distributed over the entire surface of the cell. It form cluster of cells by sticking to each other and to the surface. They even help some pathogens to adhere tightly to other cells causing infections.

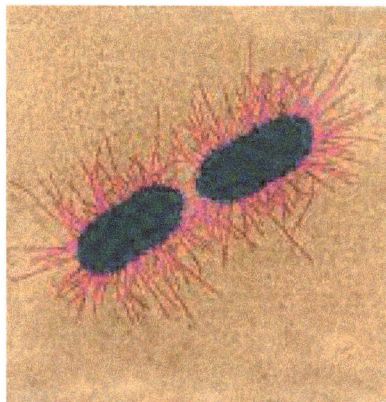

Some of the differences between fimbriae and pili are as follows:

S.N.	Characteristics	Fimbriae	Pili
1	Definition	Fimbriae are tiny bristle-like fibers arising from the surface of bacterial cells.	Pili are hair like microfibers that are thick tubular structure made up of pilin.
2	Length	Shorter than pili	Longer than fimbriae.
3	Diameter	Thin	Thicker than fimbriae.
4	Number	No. of fimbriae are 200-400 per cell.	No of pili are less 1-10 per cell.
5	Made up of	Fimbrillin protein.	Pilin protein.
6	Rigidity	Less rigid.	More rigid than fimbriae.
7	Found in	Both gram positive and gram negative bacteria.	Only gram negative bacteria.

8	Formation	Is governed by bacterial genes in the nucleoid region.	Is governed by plasmid genes.
9	Function	Responsible for cell to surface attachment. Specialized for attachment i.e. enable the cell to adhere the surfaces of other bacteria.	Responsible for bacterial conjugation. Two basic function of pili. They are gene transfer and attachment.
10	Motility	Do not function in active motility.	Type IV pili shows twitching type of motility.
11	Receptors	No receptors of other.	Serve as receptor for certain viruses.
12	Examples	Salmonella typhimurium, Shigella dysenteriae. Shigella dysenteriae uses its fimbriae to attach to the intestine and then produces a toxin that causes diarrhea.	Escherichia coli, Neisseria gonorrhoeae. Neisseria gonorrhoeae, the cause of gonorrhea, uses pili to attach to the urogenital and cervical epithelium when it causes disease.

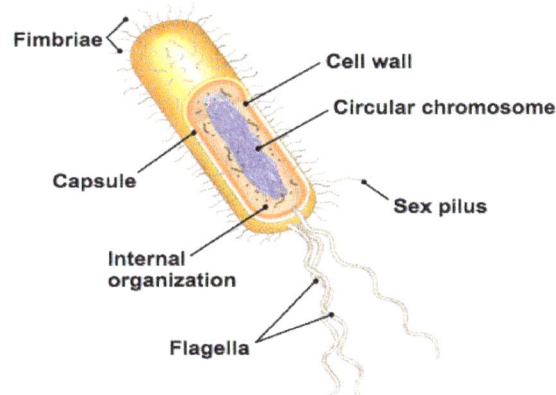

S-layer

S-layers are found in many types of bacteria, including Gram-positive and Gram-negative bacteria, cyanobacteria and the Archaea. They are composed of crystalline arrays of protein or glycoprotein subunits. In Gram-positive organisms the S-layer lies on the outer surface of the peptidoglycan layer, while in Gram-negative bacteria it is present on the outer surface of the outer membrane. It is probable that the main functions of the S-layer are to protect the bacterial cell from hostile external environments, and to act as a permeability barrier to macromolecules. To achieve the selective permeability of the S-layer, pores are present within its protein subunits to facilitate the movement of various molecules.

The presence of pores in S-layers can be demonstrated by means of atomic force microscopy which, unlike electron microscopy, does not operate under a vacuum. This allows physiological events to be monitored and by using rapid freezing techniques certain events can be 'captured'. Using AFM the surfaces of biological structures can be viewed directly, not under a vacuum, in physiologically buffered solutions. The resolution achievable with AFM is very high; for example it can reveal the pattern of protein subunits of the S-layer on the surface of the radio-tolerant bacterium Deinococcus radiodurans. By observing the inner surface of the S-layer of D. radiodurans with AFM, Müller et al. were able to demonstrate the presence of pores within the protein subunits. These pores could be seen in 'open' and 'closed' conformations, suggesting a role in regulating the passage of macromolecules into the bacterial cell.

Glycocalyx

The glycocalyx is a carbohydrate-enriched coating that covers the outside of many eukaryotic cells and prokaryotic cells, particularly bacteria . When on eukaryotic cells the glycocalyx can be a factor used for the recognition of the cell. On bacterial cells, the glycocalyx provides a protective coat from host factors. The possession of a glycocalyx on bacteria is associated with the ability of the bacteria to establish an infection.

The glycocalyx of bacteria can assume several forms. If in a condensed form that is relatively tightly associated with the underlying cell wall, the glycocalyx is referred to as a capsule. A more loosely attached glycocalyx that can be removed from the cell more easily is referred to as a slime layer.

The bacterial glycocalyx can vary in structure from bacteria to bacteria. Even particular bacteria can be capable of producing a glycocalyx of varying structure, depending upon the growth conditions and nutrients available. Generally, the glycocalyx is constructed of one or more sugars that are called saccharides. If more than one saccharide is present, the glycocalyx is described as being made of polysaccharide. In some glycocalyces, protein can also be present.

There are two prominent functions of the glycocalyx. The first function is to enable bacteria to become harder for the immune cells called phagocytes so surround and engulf. This is because the presence of a glycocalyx increases the effective diameter of a bacterium and also covers up components of the bacterium that the immune system would detect and be stimulated by. Thus, in a sense, a bacterium with a glycocalyx becomes more invisible to the immune system of a host.

Infectious strains of bacteria such as Staphylococcus, Streptococcus, and Pseudomonas tend to elaborate more glycocalyx than their corresponding non-infectious counterparts.

The second function of a bacterial glycocalyx is to promote the adhesion of the bacteria to living and inert surfaces and the subsequent formation of adherent, glycocalyx-enclosed populations that are called biofilms . Biofilm bacteria can become very hard to kill, party due to the presence of the glycocalyx material. Many persistent infections in the body are caused by bacterial biofilms. One example is the dental plaque formed by glycocalyx-producing Streptococcus mutans, which can become a focus for tooth enamel-digesting acid formed by the bacteria. Another example is the chronic lung infections formed in those afflicted with certain forms of cystic fibrosis by glycocalyx-producing Pseudomonas aeruginosa. The latter infections can cause sufficient lung damage to prove lethal.

In Vascular Endothelial Tissue

The glycocalyx is located on the apical surface of vascular endothelial cells which line the lumen. When vessels are stained with cationic dyes such as Alcian blue stain, transmission electron microscopy shows a small, irregularly shaped layer extending approximately 50–100 nm into the lumen of a blood vessel. Another study used cryotransmission electron microscopy and showed that the endothelial glycocalyx could be up to 11 μm thick. It is present throughout a diverse range of microvascular beds (capillaries) and macrovessels (arteries and veins). The glycocalyx also consists of a wide range of enzymes and proteins that regulate leukocyte and thrombocyte adherence, since its principal role in the vasculature is to maintain plasma and vessel-wall homeostasis. These enzymes and proteins include:

- Endothelial nitric oxide synthase (endothelial NOS)

- Extracellular superoxide dismutase (SOD3)

- Angiotensin converting enzyme

- Antithrombin-III

- Lipoprotein lipase

- Apolipoproteins

- Growth factors

- Chemokines

The enzymes and proteins listed above serve to reinforce the glycocalyx barrier against vascular and other diseases. Another main function of the glycocalyx within the vascular endothelium is that it shields the vascular walls from direct exposure to blood flow, while serving as a vascular permeability barrier. Its protective functions are universal throughout the vascular system, but its relative importance varies depending on its exact location in the vasculature. In microvascular tissue, the glycocalyx serves as a vascular permeability barrier by inhibiting coagulation and leukocyte adhesion. Leukocytes must not stick to the vascular wall because they are important components of the immune system that must be able to travel to a specific region of the body when needed. In arterial vascular tissue, the glycocalyx also inhibits coagulation and leukocyte adhesion, but through mediation of shear stress-induced nitric oxide release. Another protective function throughout the cardiovascular system is its ability to affect the filtration of interstitial fluid from capillaries into the interstitial space.

The glycocalyx, which is located on the apical surface of endothelial cells, is composed of a negatively charged network of proteoglycans, glycoproteins, and glycolipids.

Disruption and Disease

Because the glycocalyx is so prominent throughout the cardiovascular system, disruption to this structure has detrimental effects that can cause disease. Certain stimuli that cause atheroma may lead to enhanced sensitivity of vasculature. Initial dysfunction of the glycocalyx can be caused by hyperglycemia or oxidized low-density lipoproteins (LDLs), which then causes atherothrombosis.

In microvasculature, dysfunction of the glycocalyx leads to internal fluid imbalance, and potentially edema. In arterial vascular tissue, glycocalyx disruption causes inflammation and atherothrombosis.

Experiments have been performed to test precisely how the glycocalyx can be altered or damaged. One particular study used an isolated perfused heart model designed to facilitate detection of the state of the vascular barrier portion, and sought to cause insult-induced shedding of the glycocalyx to ascertain the cause-and-effect relationship between glycocalyx shedding and vascular permeability. Hypoxic perfusion of the glycocalyx was thought to be sufficient to initiate a degradation mechanism of the endothelial barrier. The study found that flow of oxygen throughout the blood vessels did not have to be completely absent (ischemic hypoxia), but that minimal levels of oxygen were sufficient to cause the degradation. Shedding of the glycocalyx can be triggered by inflammatory stimuli, such as tumor necrosis factor-alpha. Whatever the stimulus is, however, shedding of the glycocalyx leads to a drastic increase in vascular permeability. Vascular walls being permeable is disadvantageous, since that would enable passage of some macromolecules or other harmful antigens.

Fluid shear stress is also a potential problem if the glycocalyx is degraded for any reason. This type of frictional stress is caused by the movement of viscous fluid (i.e. blood) along the lumen boundary. Another similar experiment was carried out to determine what kinds of stimuli cause fluid shear stress. The initial measurement was taken with intravital microscopy, which showed a slow-moving plasma layer, the glycocalyx, of 1 μm thick. Light dye damaged the glycocalyx minimally, but that small change increased capillary hematocrit. Thus, fluorescence light microscopy should not be used to study the glycocalyx because that particular method uses a dye. The glycocalyx can also be reduced in thickness when treated with oxidized LDL. These stimuli, along with many other factors, can cause damage to the delicate glycocalyx. These studies are evidence that the glycocalyx plays a crucial role in cardiovascular system health.

In Bacteria and Nature

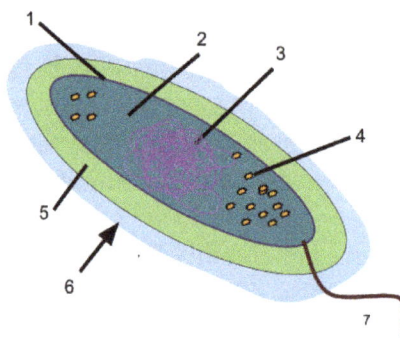

The glycocalyx exists in bacteria as either a capsule or a slime layer. Item 6 points at the glycocalyx. The difference between a capsule and a slime layer is that in a capsule polysaccharides are firmly attached to the cell wall, while in a slime layer, the glycoproteins are loosely attached to the cell wall.

A glycocalyx, literally meaning "sugar coat" (*glykys* = sweet, *kalyx* = husk), is a network of polysaccharides that project from cellular surfaces of bacteria, which classifies it as a universal surface component of a bacterial cell, found just outside the bacterial cell wall. A distinct, gelatinous glycocalyx is called a capsule, whereas an irregular, diffuse layer is called a slime layer. This coat is extremely hydrated and stains with ruthenium red.

Bacteria growing in natural ecosystems, such as in soil, bovine intestines, or the human urinary tract, are surrounded by some sort of glycocalyx-enclosed microcolony. It serves to protect the bacterium from harmful phagocytes by creating capsules or allowing the bacterium to attach itself to inert surfaces, such as teeth or rocks, via biofilms (e.g. *Streptococcus pneumoniae* attaches itself to either lung cells, prokaryotes, or other bacteria which can fuse their glycocalices to envelop the colony).

In the Digestive Tract

A glycocalyx can also be found on the apical portion of microvilli within the digestive tract, especially within the small intestine. It creates a meshwork 0.3 μm thick and consists of acidic mucopolysaccharides and glycoproteins that project from the apical plasma membrane of epithelial absorptive cells. It provides additional surface for adsorption and includes enzymes secreted by the absorptive cells that are essential for the final steps of digestion of proteins and sugars.

Other Generalized Functions

- Protection: Cushions the plasma membrane and protects it from chemical injury

- Immunity to infection: Enables the immune system to recognize and selectively attack foreign organisms

- Defense against cancer: Changes in the glycocalyx of cancerous cells enable the immune system to recognize and destroy them.

- Transplant compatibility: Forms the basis for compatibility of blood transfusions, tissue grafts, and organ transplants

- Cell adhesion: Binds cells together so that tissues do not fall apart

- Inflammation regulation: Glycocalyx coating on endothelial walls in blood vessels prevents leukocytes from rolling/binding in healthy states.

- Fertilization: Enables sperm to recognize and bind to eggs

- Embryonic development: Guides embryonic cells to their destinations in the body

Flagella

A flagellum (plural, flagella) is a long, whip-like projection or appendage of a cell composed of microtubules (long, slender, proteintubes) and used in motility. They help propel cells and organisms in a whip-like motion. The flagellum of eukaryotes usually moves with an "S" motion and is surrounded by cell membrane.

Eukaryote flagella are similar to cilia—another structure that extends out from the surface of cell and is used for movement—in that both are composed of nine pairs of microtubules (nine microtubule doublets) arranged around its circumference and one pair of microtubules (two microtubule siglets) running down the center, the 9 + 2 structure. However, flagella are longer and typically occur singly or in pairs, or at least much smaller numbers than cilia, which occur in large numbers. There are also functional differences in terms of type of movement or force exerted. Flagella use a whip-like action to create movement of the whole cell, such as the movement of sperm in the reproductive tract. Cilia primarily use a waving action to move substances across the cell, such as the ciliary esculator found in the respiratory tract. Cilia may also function as sensory organs.

Prokaryotes may have one or many flagella for locomotion, but these differ significantly from flagella in eukaryotes. Flagella in archaebacteria are distinct from both of those types.

The structural similarity of cilia and eukaryote flagella, and the substantial differences between flagella in eukaryotes and prokaryotes, is such that some authorities group cilia and eukaryote flagella together and consider cilium simply a special type of flagellum—one organized such that many flagella (cilia) may work in synchrony. The term undulipodium is used for an intracellular projection of a eukaryote cell with a microtuble array and includes both flagella and cilia.

In Protozoa— a diverse group of single-celled, microscopic or near microscopic protist eukaryotes that commonly show characteristics usually associated with animals—those organisms with flagella (flagellates) are generally placed in the phylum Zoomastigina (or Mastigophora), whereas those with cilia (ciliates) are placed in phylum Ciliophora.

Many parasites that affect human health or economy are flagellates. These include such parasitic protozoans as members of the genera Trypanosoma (cause of African trypanosomiasis, or sleeping sickness, and Chagas disease, or South American trypanosomiasis), Leishmania (cause of leishmania, which affects millions of people in Africa, Asia, and Latin America), and Giardia(causes giardiasis). Trypanosoma species are carried from host to host by bloodsucking invertebrates, such as the tsetse fly and conenose bugs; Leishmania is carried by sand flies; and Giardia is carried by muskrats and beavers.

Protozoan flagellates play important ecological roles in food chains as major consumers of bacteriaand other protists and the recycling of nutrients.

The flagellum has been a prominent focal point in the debate between those advocating Darwinismand those advocating intelligent design.

Overview

Bacterial flagella are entirely outside the cell membrane (plasma membrane) and are normally visible only with the aid of an electron microscope. In some bacterial species, the flagella twine together helically outside the cell body to form a bundle large enough to be visible in a light microscope. These structures are quite unrelated to the flagella of eukaryotes.

A eukaryote cell usually only has about one or two flagella. The flagella also may have hair ormastigonemes, scales, connecting membranes, and internal rods. Flagellates move by whipping the flagella on the flagellate side to side. A sperm cell moves by means of a single flagellum. In a multicellular organism, cilia or flagella can also extend out from stationary cells that are held in place as part of a tail goes into a layer of tissue. In eukaryotic cells, flagella are active in movements involving feeding and sensation.

Movement of a unicellular organisms by flagella can be relatively swift, whether it be Euglena with its emergent flagellum or a sperm cell with its flagellum.

Unlike bacteria, eukaryote flagella have an internal structure comprised of nine doublets of microtubules forming a cylinder around a central pair of microtubules. The peripheral doublets are linked to each other by proteins. These proteins include dynein, a molecular motor that can cause flagella to bend, and propel the cell relative to its environment or propel water or mucus relative to the cell.

The three major domains of organisms (as classified in the three domain system) each have different structural/functional aspects of the flagella:

- Bacterial flagella are helical filaments that rotate like screws.

- Archaeal (archaebacterial) flagella are superficially similar, but are different in many details and considered non-homologous.

- Eukaryotic flagella—those of animal, plant, and protist cells— are complex cellular projections that lash back and forth.

Bacterial Flagellum

The bacterial filament is composed of the protein flagellin and is a hollow tube 20 nanometers thick. It is helical, and has a sharp bend just outside the outer membrane called the "hook" which allows the helix to point directly away from the cell. A shaft runs between the hook and the basal body, passing through protein rings in the cell's membranes that act as bearings. Gram-positive organisms have 2 basal body rings, one in the peptidoglycan layer and one in the plasma membrane. Gram-negative organisms have 4 rings: L ring associates with the lipopolysaccharides, P ring associates with peptidoglycan layer, M ring embedded in the plasma membrane, and the S ring directly attached to the plasma membrane. The filament ends with a capping protein.

The bacterial flagellum is driven by a rotary engine composed of protein, located at the flagellum's anchor point on the inner cell membrane. The engine is powered by proton motive force, i.e., by the flow of protons (i.e., hydrogen ions) across the bacterial cell membrane due to a concentration gradient set up by the cell's metabolism. (In Vibrio species, the motor is a sodium ion pump, rather than a proton pump). The rotor transports protons across the membrane and is turned in the process. The rotor by itself can operate at 6,000 to 17,000 revolutions per minute (rpm), but with a filament attached usually only reaches 200 to 1000 rpm.

The components of the flagellum are capable of self-assembly in which the component proteins associate spontaneously without the aid of enzymes or other factors. Both the basal body and the filament have a hollow core, through which the component proteins of the flagellum are able to move into their respective positions. The filament grows at its tip rather than at the base. The basal body has many traits in common with some types of secretory pores, which have a hollow rod-like "plug" in their centers extending out through the plasma membrane. It was thought that bacterial flagella may have evolved from such pores, though it is now considered that these pores are derived from flagella.

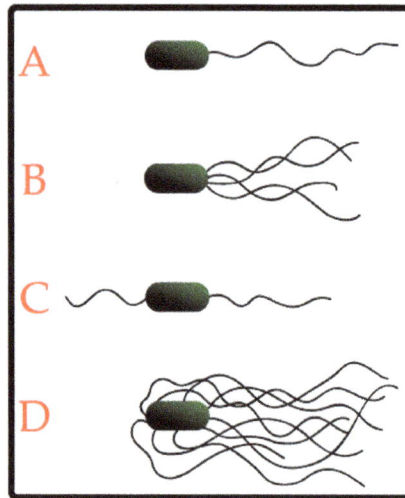

Examples of bacterial flagella arrangement schemes. A-Monotrichous;
B-Lophotrichous; C-Amphitrichous; D-Peritrichous;

Different species of bacteria have different numbers and arrangements of flagella. Monotrichous bacteria have a single flagellum (example:Vibrio cholerae). Lophotrichous bacteria have multiple flagella located at the same spot on the bacteria's surface, which act in concert to drive the bacteria in a single direction. Amphitrichous bacteria have a single flagellum each on two opposite ends. (Only one end's flagellum operates at a time, allowing the bacteria to reverse course rapidly by switching which flagellum is active.) Peritrichous bacteria have flagella projecting in all directions (example: Escherichia coli).

Some species of bacteria (those of Spirochete body form) have a specialized type of flagellum called axial filament that is located in the periplasmic space, the rotation of which causes the entire bacterium to corkscrew through its usually viscous medium.

Anticlockwise rotation of monotrichous polar flagella thrusts the cell forward with the flagellum trailing behind. Periodically, the direction of rotation is briefly reversed, causing what is known as a "tumble," and results in reorientation of the cell. The direction at the end of the tumble state is random. The length of the run state is extended when the bacteria moves through a favorable gradient.

Archaeal Flagellum

The archaeal flagellum is superficially similar to the bacterial (or eubacterial) flagellum; in the 1980s they were thought to be homologous on the basis of gross morphology and behavior. Both flagella consist of filaments extending outside of the cell and rotate to propel the cell.

However, discoveries in the 1990s have revealed numerous detailed differences between the archaeal and bacterial flagella. These include:

- Bacterial flagella are powered by a flow of H^+ ions (or occasionally Na^+ ions); archaeal flagella are almost certainly powered by ATP. The torque-generating motor that powers rotation of the archaeal flagellum has not been identified.

- While bacterial cells often have many flagellar filaments, each of which rotates independently, the archaeal flagellum is composed of a bundle of many filaments that rotate as a single assembly.

- Bacterial flagella grow by the addition of flagellin subunits at the tip; archaeal flagella grow by the addition of subunits to the base.

- Bacterial flagella are thicker than archaeal flagella, and the bacterial filament has a large enough hollow "tube" inside that the flagellin that subunits can flow up the inside of the filament and get added at the tip; the archaeal flagellum is too thin to allow this.

- Many components of bacterial flagella share sequence similarity to components of the type III secretion systems, but the components of bacterial and archaeal flagella share no sequence similarity. Instead, some components of archaeal flagella share sequence and morphological similarity with components of type IV pili, which are assembled through the action of type II secretion systems. (The nomenclature of pili and protein secretion systems is not consistent.)

These differences mean that the bacterial and archaeal flagella are a classic case of biological analogy, or convergent evolution, rather than homology (sharing common origin). However, in comparison to the decades of well-publicized study of bacterial flagella (e.g. by Berg), archaeal flagella have only recently begun to get serious scientific attention. Therefore, many assume erroneously that there is only one basic kind of prokaryotic flagellum, and that archaeal flagella are homologous to it.

Eukaryotic Flagellum

The eukaryotic flagellum is completely different from the prokaryote flagella in structure and assumedly historical origin. The only shared characteristics among bacterial, archaeal, and eukaryotic flagella is their superficial appearance; they are intracellular extensions used in creating movement. Along with cilia, eukaryote flagella make up a group of organelles known as undulipodia.

A eukaryotic flagellum is a bundle of nine fused pairs of microtubule doublets surrounding two central single microtubules. The so-called 9+2 structure is characteristic of the core of the eukaryotic flagellum called an axoneme. At the base of a eukaryotic flagellum is a basal body, "blepharoplast" or kinetosome, which is the microtubule organizing center for flagellar microtubules and is about 500 nanometers long. Basal bodies are structurally identical to centrioles.

The flagellum is encased within the cell's plasma membrane, so that the interior of the flagellum is accessible to the cell's cytoplasm. Each of the outer 9 doublet microtubules extends a pair of dynein arms (an "inner" and an "outer" arm) to the adjacent microtubule; these dynein arms are responsible for flagellar beating, as the force produced by the arms causes the microtubule doublets to slide against each other and the flagellum as a whole to bend. These dynein arms produce force through ATP hydrolysis. The flagellar axoneme also contains radial spokes, polypeptide complexes extending from each of the outer 9 mictrotubule doublets towards the central pair, with the "head" of the spoke facing inwards. The radial spoke is thought to be involved in the regulation of flagellar motion, although its exact function and method of action are not yet understood.

Motile flagella serve for the propulsion of single cells (e.g. swimming of protozoa and spermatozoa) and the transport of fluids (e.g. transport of mucus by stationary flagellated cells in the trachea).

Additionally, immotile flagella are vital organelles in sensation and signal transduction across a wide variety of cell types (e.g. eye: rod photoreceptor cells, nose: olfactory receptor neurons, ear: kinocilium in cochlea).

Intraflagellar transport (IFT), the process by which axonemal subunits, transmembrane receptors, and other proteins are moved up and down the length of the flagellum, is essential for proper functioning of the flagellum, in both motility and signal transduction.

Arthropod Flagellum

In Chelicerata (an arthropod subphylum that includes spiders, scorpions, horseshow crabs, sea spiders, and so forth), the flagellum is a non-segmental, pluri-articulated whip, present in the arachnid orders Schizomida, Thelyphonida, and Palpigradi. In Schizomida, the flagellum of the male has complex morphology and is widely used in taxonomy.

Plasmid

A plasmid is a small, circular piece of DNA that is different than the chromosomal DNA, which is all the genetic material found in an organism's chromosomes. It replicates independently of chromosomal DNA. Plasmids are mainly found in bacteria, but they can also be found in archaea and multicellular organisms. Plasmids usually carry at least one gene, and many of the genes that plasmids carry are beneficial to their host organisms. Although they have separate genes from their hosts, they are not considered to be independent life.

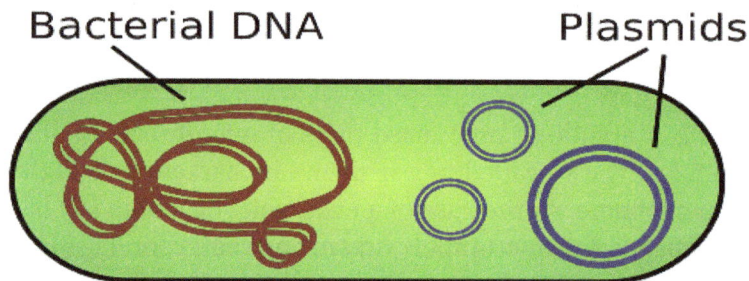

This simplified figure depicts a bacterium's chromosomal DNA in red and plasmids in blue.

Functions of Plasmids

Plasmids have many different functions. They may contain genes that enhance the survival of an organism, either by killing other organisms or by defending the host cell by producing toxins. Some plasmids facilitate the process of replication in bacteria. Since plasmids are so small, they usually only contain a few genes with a specific function (as opposed to a large amount of non-coding DNA). Multiple plasmids can coexist in the same cell, each with different functions. The functions are further detailed in the section "Specific Types of Plasmids" below.

General Types of Plasmids

Conjugative and Non-conjugative

There are many ways to classify plasmids from general to specific. One way is by grouping them as either conjugative or non-conjugative. Bacteria reproduce by sexual conjugation, which is the transfer of genetic material from one bacterial cell to another, either through direct contact or a bridge between the two cells. Some plasmids contain genes called transfer genes that facilitate the beginning of conjugation. Non-conjugative plasmids cannot start the conjugation process, and they can only be transferred through sexual conjugation with the help of conjugative plasmids.

Incompatibility

Another plasmid classification is by incompatibility group. In a bacterium, different plasmids can only co-occur if they are compatible with each other. An incompatible plasmid will be expelled from the bacterial cell. Plasmids are incompatible if they have the same reproduction strategy in the cell; this allows the plasmids to inhabit a certain territory within it without other plasmids interfering.

Specific Types of Plasmids

There are five main types of plasmids: fertility F-plasmids, resistance plasmids, virulence plasmids, degradative plasmids, and Col plasmids.

Fertility F-plasmids

Fertility plasmids, also known as F-plasmids, contain transfer genes that allow genes to be transferred from one bacteria to another through conjugation. These make up the broad category of conjugative plasmids. F-plasmids are episomes, which are plasmids that can be inserted into chromosomal DNA. Bacteria that have the F-plasmid are known as F positive (F$^+$), and bacteria without it are F negative (F$^-$). When an F$^+$ bacterium conjugates with an F$^-$ bacterium, two F$^+$ bacterium result. There can only be one F-plasmid in each bacterium.

Resistance Plasmids

Resistance or R plasmids contain genes that help a bacterial cell defend against environmental factors such as poisons or antibiotics. Some resistance plasmids can transfer themselves through conjugation. When this happens, a strain of bacteria can become resistant to antibiotics. Recently, the type bacterium that causes the sexually transmitted infection gonorrhea has become so resistant to a class of antibiotics called quinolones that a new class of antibiotics, called cephalosporins, has started to be recommended by the World Health Organization instead. The bacteria may even become resistant to these antibiotics within five years. According to NPR, overuse of antibiotics to treat other infections, like urinary tract infections, may lead to the proliferation of drug-resistant strains.

Virulence Plasmids

When a virulence plasmid is inside a bacterium, it turns that bacterium into a pathogen, which is an agent of disease. Bacteria that cause disease can be easily spread and replicated among affected individuals. The bacterium Escherichia coli (E. coli) has several virulence plasmids. E. coli is found naturally in the human gut and in other animals, but certain strains of E. coli can cause severe diarrhea and vomiting. Salmonella enterica is another bacterium that contains virulence plasmids.

Degradative Plasmids

Degradative plasmids help the host bacterium to digest compounds that are not commonly found in nature, such as camphor, xylene, toluene, and salicylic acid. These plasmids contain genes for special enzymes that break down specific compounds. Degradative plasmids are conjugative.

Col Plasmids

Col plasmids contain genes that make bacteriocins (also known as colicins), which are proteins that kill other bacteria and thus defend the host bacterium. Bacteriocins are found in many types of bacteria including E. coli, which gets them from the plasmid ColE1.

Applications of Plasmids

Humans have developed many uses for plasmids and have created software to record the DNA sequences of plasmids for use in many different techniques. Plasmids are used in genetic engineering to amplify, or produce many copies of, certain genes. In molecular cloning, a plasmid is a type of vector. A vector is a DNA sequence that can transport foreign genetic material from one cell to another cell, where the genes can be further expressed and replicated. Plasmids are useful in cloning short segments of DNA. Also, plasmids can be used to replicate proteins, such as the protein that codes for insulin, in large amounts. Additionally, plasmids are being investigated as a way to transfer genes into human cells as part of gene therapy. Cells may lack a specific protein if the patient has a hereditary disorder involving a gene mutation. Inserting a plasmid into DNA would allow cells to express a protein that they are lacking.

Ribosome

While examining the animal and plant cell through a microscope, you might have seen numerous organelles that work together to complete the cell activities. One of the essential cell organelles are ribosomes, which are in charge of protein synthesis. The ribosome is a complex made of protein and RNA and which adds up to numerous million Daltons in size and assumes an important part in the course of decoding the genetic message reserved in the genome into protein.

The essential chemical step of protein synthesis is peptidyl transfer, that the developing or nascent peptide is moved from one tRNA molecule to the amino acid together with another tRNA. Amino acids are included in the developing polypeptide in line with the arrangement of codons of a mRNA. The ribosome, therefore, has necessary sites for one mRNA and no less than two tRNAs.

Made of two subunits, the big and the little subunit which comprises a couple of ribosomal RNA (rRNA) molecules and an irregular number of ribosomal proteins. Numerous protein factors catalyze distinct impression of protein synthesis. The translation of the genetic code is of essential significance for the manufacturing of useful proteins and for the growth of the cell.

Structure

Ribosomes are made of proteins and ribonucleic acid (abbreviated as RNA), in almost equal amounts. It comprises of two sections, known as subunits. The tinier subunit is the place the mRNA binds and it decodes, whereas the bigger subunit is the place the amino acids are included.

Both subunits comprise of both ribonucleic acid and protein components and are linked to each other by interactions between the proteins in one subunit and the rRNAs in the other subunit. The ribonucleic acid is obtained from the nucleolus, at the point where ribosomes are arranged in a cell.

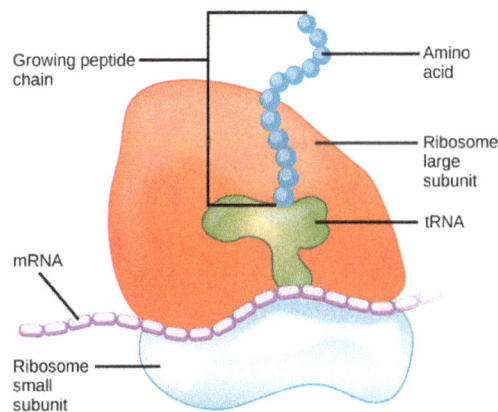

The structures of ribosomes include:

- Situated in two areas of the cytoplasm.

- They are seen scattered in the cytoplasm and a few are connected to the endoplasmic reticulum.

- Whenever joined to the ER they are called the rough endoplasmic reticulum.

- The free and the bound ribosomes are very much alike in structure and are associated with protein synthesis.

- Around 37 to 62% of RNA is comprised of RNA and the rest is proteins.

- Prokaryotes have 70S ribosomes respectively subunits comprising the little subunit of 30S and the bigger subunit of 50S. Eukaryotes have 80S ribosomes respectively comprising of little (40S) and substantial (60S) subunits.

- The ribosomes seen in the chloroplasts of mitochondria of eukaryotes are comprised of big and little subunits composed of proteins inside a 70S particle.

- Share a center structure which is very much alike to all ribosomes in spite of changes in its size.

- The RNA is arranged in different tertiary structures. The RNA in the bigger ribosomes is into numerous continuous infusions as they create loops out of the center of the structure without disturbing or altering it.

- The contrast between those of eukaryotic and bacteria are utilized to make antibiotics that can crush bacterial disease without damaging human cells.

Ribosomes Size

Ribosomes comprise of two subunits that are suitably composed and function as one to translate the mRNA into a polypeptide chain amid protein synthesis. Due to the fact that they are made from two subunits of differing size, they are a little longer in the hinge than in diameter. They vary in size between prokaryotic cells and eukaryotic cells.

The prokaryotic is comprised of a 30s (Svedberg) subunit and a 50s (Svedberg) subunit meaning 70s for the entire organelle equal to the molecular weight of 2.7×10^6 Daltons. Prokaryotic ribosomes are about 20 nm (200 Å) in diameter and are made of 35% ribosomal proteins and 65% rRNA.

Notwithstanding, the eukaryotic are amidst 25 and 30 nm (250–300 Å) in diameter. They comprise of a 40s (Svedberg) subunit and a 60s (Svedberg) subunit which means 80s (Svedberg) for the entire organelle which is equal to the molecular weight of 4×10^6 Daltons.

Location

Ribosomes are organelles located inside the animal, human cell, and plant cells. They are situated in the cytosol, some bound and free-floating to the membrane of the coarse endoplasmic reticulum.

They are utilized in decoding DNA (deoxyribonucleic acid) to proteins and no rRNA is forever bound to the RER, they release or bind as directed by the kind of protein they proceed to combine. In an animal or human cell, there could be up to 10 million ribosomes and numerous ribosomes can be connected to the equivalent mRNA strand, this structure is known as a POLYSOME.

Function

When it comes to the main functions of ribosomes, they assume the role of bringing together amino acids to form particular proteins, which are important for completing the cell's activities.

Protein is required for numerous cell functions, for example, directing chemical processes or fixing the damage. Ribosomes can yet be discovered floating inside the cytoplasm or joined to the endoplasmic reticulum.

The other functions include:

1. The procedure of creation of proteins, the deoxyribonucleic acid makes mRNA by the step of DNA transcription.

2. The hereditary information from the mRNA is converted into proteins amid DNA translation.

3. The arrangements of protein assembly amid protein synthesis are indicated in the mRNA.

4. The mRNA is arranged in the nucleus and is moved to the cytoplasm for an additional operation of protein synthesis.

5. The proteins which are arranged by the ribosomes currently in the cytoplasm are utilized inside the cytoplasm by itself. The proteins created by the bound ribosomes are moved outside the cell.

Taking into consideration their main function in developing proteins, it is clear that a cell can't function in the absence of ribosomes.

Those that live inside bacteria, parasites and different creatures, for example, lower and microscopic level creatures are the ones which are called prokaryotic ribosomes. While those that live inside humans and others such as higher level creatures are those ones we call the eukaryotic ribosome.

The other major differences include:

1. Prokaryotes have 70S ribosomes, singly made of a 30S and a 50S subunit. While the Eukaryotes have 80S ribosomes, singly made of a 40S and 60S subunit.

2. 70S Ribosomes are relatively smaller than 80S while the 80S Ribosomes are relatively bigger than 70S ribosomes.

3. Prokaryotes have 30S subunit with a 16S RNA subunit and comprise of 1540 nucleotides bound to 21 proteins. The 50S subunit gets produced from a 5S RNA subunit that involves 120 nucleotides, a 23S RNA subunit that contains 2900 nucleotides and 31 proteins.

4. Eukaryotes have 40S subunit with 18S RNA and also 33 proteins and 1900 nucleotides. The big subunit contains 5S RNA and also 120 nucleotides, 4700 nucleotides and also 28S RNA, 5.8S RNA as well as 160 nucleotides subunits and 46 proteins.

5. Eukaryotic cells have mitochondria and chloroplasts as organelles and those organelles additionally have ribosomes 70S. Hence, eukaryotic cells have different kinds of ribosomes (70S and 80S), while prokaryotic cells just have 70S ribosomes.

Mesosomes

Mesosomes are the invaginated structures formed by the localized infoldings of the plasma membrane. The invaginated structures comprise of vesicles, tubules of lamellar whorls.

Generally mesosomes are found in association with nuclear area or near the site of cell division. They are absent in eukaryotes. The lamellae are formed by flat vesicles when arranged parallely. Some of the lamellae are connected to the cell membrane. The lamellar whorl can be observed in Nitrobacter, Nitro monas and Nitrococcus.

The vesicles are formed probably by invagination and tubular accretion of the plasma membrane. The structure of vesicle becomes interrupted due to constriction at equal distance. The constriction does not cause the complete separation of tubules. Closely packed spherical vesicles are seen in Chromatium and Rhodospirillum rubrum.

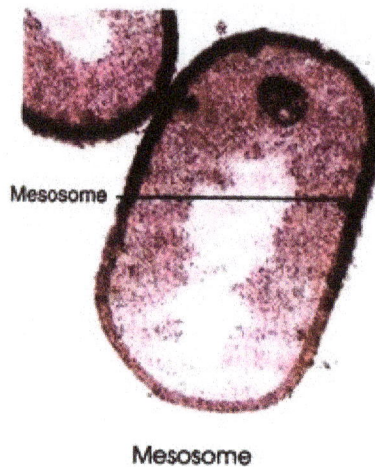

Mesosome

In some purple bacteria the vesicular bodies are flattened and stacked into the regular plates like thylakoids. Salton and Owen have suggested that the mesosomes are formed due to vesicularization of outer half of the lipid bilayer.

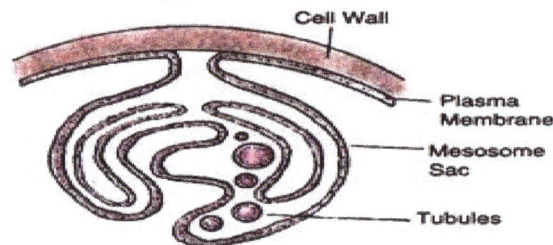

Fig. 4.15 : The bacterial mesosome
(diagrammatic)

However, they are the special cell membrane components, the proteins of which differ from the cell membrane. The exact structure and function of mesosomes are not known. However, it has been suggested that these are artifacts (i.e. a structure that appears in microscopic preparations due to the method of preparation).

Moreover, mesosomes are supposed to take part in respiration but they are not analogous to mitochondria because they lack outer membrane. Respiratory enzymes have been found to be present in cell membrane.

In the vesicle of mesosomes the respiratory enzymes and the components of electron transport such as ATPase, dehydrogenase, cytochrome are either absent or present in low amount. This emphasizes their inability to carryout transport process in which the membrane is energised. In addition, mesosomes are supposed as a site for synthesis of some of wall membranes.

Mesosomes might play a role in reproduction also. During binary fission a cross wall is formed resulting in formation of two cells. Mesosomes begin the formation of septum and attach bacterial DNA to the cell membrane. It separates the bacterial DNA into each daughter cell. In addition, the infoldings of mesosomes increase the surface area of plasma membrane that in turn increases the absorption of nutrients.

Gas Vacuoles

Gas vacuoles are aggregates of hollow cylindrical structures called gas vesicles. They are located inside some bacteria. A membrane that is permeable to gas bound each gas vesicle. The inflation and deflation of the vesicles provides buoyancy, allowing the bacterium to float at a desired depth in the water.

Bacteria that are known as cyanobacteria contain gas vacuoles. Cyanobacteria, which used to be called blue-green algae , live in water and manufacture their own food from the photosynthetic energy of sunlight. Studies have demonstrated that the inflation and deflation of the gas vesicles is coordinated with the light. The buoyancy provided by the gas vacuoles enables the bacteria to float near the surface during the day to take advantage of the presence of sunlight for the manufacture of food, and to sink deeper at night to harvest nutrients that have sunk down into the water.

Gas vesicles are also found in some archae, bacteria that are thought to have branched off from a common ancestor of eukaryotes and prokaryotes at a very early stage in evolution . For example, the gas vesicles in the bacterium Halobacterium NRC-1 allow the bacteria to float in their extremely salt water environments (the bacteria are described as halophilic, or "salt loving." The detailed genetic analysis that has been done with this bacterium indicates that at least 13 to 14 genes are involved in production of the two gas vesicle structural proteins and other, perhaps regulatory, proteins. For example, some proteins may sense the environment and act to trigger synthesis of the vesicles. Vesicle synthesis is known to be triggered by low oxygen concentrations.

The gas vesicles tend to be approximately 75 nanometers in diameter. Their length is variable, ranging from 200 to 1000 nanometers, depending on the species of bacteria. The vesicles are constructed of a single small protein. In at least some vesicles these proteins are linked together by another protein. The interior of the protein shell is very hydrophobic (water-hating), so that water is excluded from the inside of the vesicles. Yet it is still unclear how the regular arrangement of proteins produces a shell that is permeable to gas. Presumably there must be enough space in between the protein subunits to permit the passage of air.

Bacterial Microcompartments

Bacterial microcompartments are large supramolecular assemblies, resembling viruses in size and shape, found inside many bacterial cells. A protein-based shell encapsulates a series of sequentially acting enzymes in order to sequester certain sensitive metabolic processes within the cell. Crystal structures of the individual shell proteins have revealed details about how they self-assemble and how pores through their centers facilitate molecular transport into and out of the microcompartments. Biochemical and genetic studies have shown that enzymes are directed to the interior in some cases by special targeting sequences in their termini. Together, these findings open up prospects for engineering bacterial microcompartmentswith novel functionalities for applications ranging from metabolic engineering to targeted drug delivery.

Shells

Protein Families Forming the Shell

The BMC shell appears icosahedral or quasi-icosahedral, and is formed by (pseudo)hexameric and pentameric protein subunits.

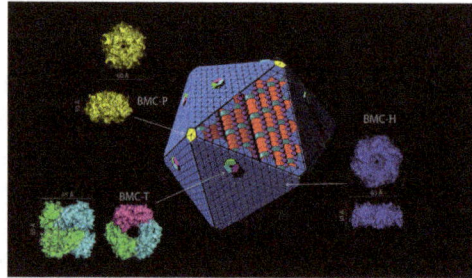

The three types of proteins (BMC-H, BMC-T and BMC-P) known to form
the shell of BMCs. The encapsulated enzymes/proteins (shown in purple,
red and turquoise) constitute a metabolic reaction sequence.

The BMC Shell Protein Family

The major constituents of the BMC shell are proteins containing Pfam00936 domain(s). These proteins form oligomers that are hexagonal in shape and are thought to form the facets of the shell.

Single-domain Proteins (BMC-H)

The BMC-H proteins, which contain a single copy of the Pfam00936 domain, are the most abundant component of the facets of the shell. The crystal structures of a number of these proteins have been determined, showing that they assemble into cyclical hexamers, typically with a small pore in the center. This opening is proposed to be involved in the selective transport of the small metabolites across the shell.

Tandem-domain Proteins (BMC-T)

A subset of shell proteins are composed of tandem (fused) copies of the Pfam00936 domain (BMC-T proteins). The structurally characterized BMC-T proteins form trimers that are pseudohexameric in shape. Some BMC-T crystal structures show that the trimers can stack in a face-to-face fashion. In such structures, one pore from one trimer is in an "open" conformation, while the other is closed – suggesting that there may be an airlock-like mechanism that modulates the permeability of some BMC shells. Another subset of BMC-T proteins contains a cluster, and may be involved in electron transport across the BMC shell.

The EutN/CcmL Family (BMC-P)

Twelve pentagonal units are necessary to cap the vertices of an icosahedral shell. Crystal structures of proteins from the EutN/CcmL family (Pfam03319) have been solved and they typically form pentamers (BMC-P). The importance of the BMC-P proteins in shell formation seems to vary among the different BMCs. It was shown that they are necessary for the formation of the shell of the PDU BMC as mutants in which the gene for the BMC-P protein was deleted cannot form shells, but not for the alpha-carboxysome: without BMC-P proteins, carboxysomes will still assemble and many are elongated; these mutant carboxysomes appear to be "leaky".

Origin of BMC and Relation to Viral Capsids

While the BMC shell is architecturally similar to many viral capsids, the shell proteins have not been found to have any structural or sequence homology to capsid proteins. Instead, structural and sequence comparisons suggest that both BMC-H (and BMC-T) and BMC-P, most likely, have evolved from bona fide cellular proteins, namely, PII signaling protein and OB-fold domain-containing protein, respectively.

Permeability of the Shell

It is well established that enzymes are packaged within the BMC shell and that some degree of metabolite and cofactor sequestration must occur. However, other metabolites and cofactors must also be allowed to cross the shell in order for BMCs to function. For example, in carboxysomes, ribulose-1,5-bisphosphate, bicarbonate, and phosphoglycerate must cross the shell, while carbon dioxide and oxygen diffusion is apparently limited. Similarly, for the PDU BMC, the shell must be permeable to propanediol, propanol, propionyl-phosphate, and potentially also vitamin B12, but it is clear that propionaldehyde is somehow sequestered to prevent cell damage. There is some evidence that ATP must also cross some BMC shells.

It has been proposed that the central pore formed in the hexagonal protein tiles of the shell are the conduits through which metabolites diffuse into the shell. For example, the pores in the carboxysome shell have an overall positive charge, which has been proposed to attract negatively charged substrates such as bicarbonate. In the PDU microcompartment, mutagenesis experiments have shown that the pore of the PduA shell protein is the route for entry of the propanediol substrate. For larger metabolites, a gating mechanism in some BMC-T proteins is apparent. In the EUT microcompartment, gating of the large pore in the EutL shell protein is regulated by the presence of the main metabolic substrate, ethanolamine.

The presence of iron-sulfur clusters in some shell proteins, presumably in the central pore, has led to the suggestion that they can serve as a conduit through which electrons can be shuttled across the shell.

Types

A recent comprehensive survey of microbial genome sequence data indicated up to ten different metabolic functions encapsulated by BMC shells. The majority are involved in either carbon fixation (carboxysomes) or aldehyde oxidation (metabolosomes).

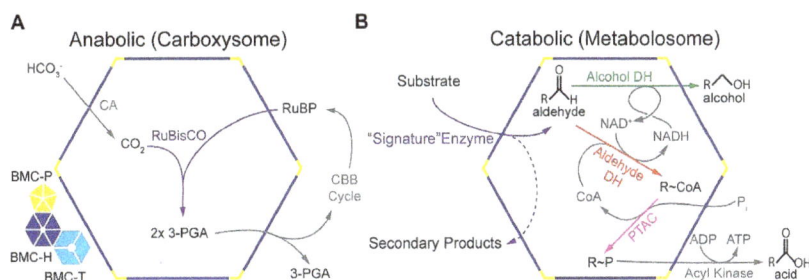

Generalized function schematic for experimentally characterized BMCs. (A) Carboxysome. (B) Metabolosome. Reactions in gray are peripheral reactions to the core BMC chemistry. BMC shell protein oligomers are depicted on the left: blue, BMC-H; cyan, BMC-T; yellow, BMC-P. 3-PGA, 3-phosphoglycerate, and RuBP, ribulose 1,5-bisphosphate.

Carboxysomes: Carbon Fixation

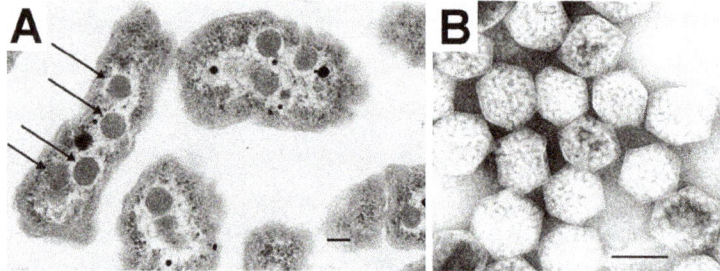

Electron micrographs showing alpha-carboxysomes from the chemoautotrophic bacterium *Halothiobacillus neapolitanus*: (A) arranged within the cell, and (B) intact upon isolation. Scale bars indicate 100 nm.

Carboxysomes encapsulate ribulose-1, 5-bisphosphate carboxylase/oxygenase (RuBisCO) and carbonic anhydrase in carbon-fixing bacteria as part of a carbon concentrating mechanism. Bicarbonate is pumped into the cytosol and diffuses into the carboxysome, where carbonic anhydrase converts it to carbon dioxide, the substrate of RuBisCO. The carboxysome shell is thought to be only sparingly permeable to carbon dioxide, which results in an effective increase in carbon dioxide concentration around RuBisCO, thus enhancing carbon fixation. Mutants that lack genes coding for the carboxysome shell display a high carbon requiring phenotype due to the loss of the concentration of carbon dioxide, resulting in increased oxygen fixation by RuBisCO. The shells have also been proposed to restrict the diffusion of oxygen, thus preventing the oxygenase reaction, reducing wasteful photorespiration.

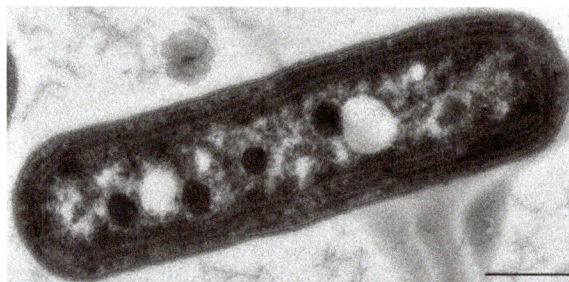

Electron micrograph of Synechococcus elongatus PCC 7942 cell showing the carboxysomes as polyhedral dark structures. Scale bar indicates 500 nm.

Metabolosomes: Aldehyde Oxidation

In addition to the anabolic carboxysomes, several catabolic BMCs have been characterized that participate in the heterotrophic metabolism via short-chain aldehydes; they are collectively termed metabolosomes.

These BMCs share a common encapsulated chemistry driven by three core enzymes: aldehyde dehydrogenase, alcohol dehydrogenase, and phosphotransacylase. Because aldehydes can be toxic to cellsand/or volatile, they are thought to be sequestered within the metabolosome. The aldehyde is initially fixed to coenzyme A by a NAD^+-dependent aldehyde dehydrogenase, but these two cofactors must be recycled, as they apparently cannot cross the shell. These recycling reactions are catalyzed by an alcohol dehydrogenase (NAD^+), and a phosphotransacetylase (coenzyme A), resulting in a phosphorylated acyl compound that can readily be a source of substrate-level phosphorylation or enter central metabolism, depending on if the organism is growing aerobically or anaerobically. It seems that most, if not all, metabolosomes utilize these core enzymes. Metabolosomes also encapsulate another enzyme that is specific to the initial substrate of the BMC, that generates the aldehyde; this is considered the signature enzyme of the BMC.

PDU BMCs

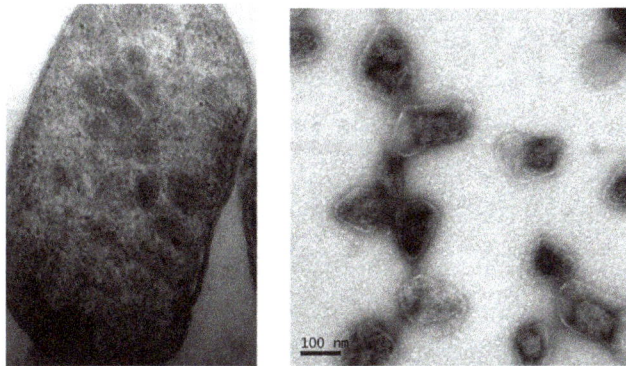

Electron micrograph of Escherichia coli cell expressing the PDU BMC genes (left), and purified PDU BMCs from the same strain (right).

Some bacteria can use 1,2-propanediol as a carbon source. They use a BMC to encapsulate several enzymes used in this pathway. The PDU BMC is typically encoded by a 21 gene locus. These genes are sufficient for assembly of the BMC since they can be transplanted from one type of bacterium to another, resulting in a functional metabolosome in the recipient. This is an example of bioengineering that likewise provides evidence in support of the selfish operon hypothesis. 1,2-propanediol is dehydrated to propionaldehyde by propanediol dehydratase, which requires vitamin B12 as a cofactor. Propionaldehyde causes DNA mutations and as a result is toxic to cells, possibly explaining why this compound is sequestered within a BMC. The end-products of the PDU BMC are propanol and propionyl-phosphate, which is then dephosphorylated to propionate, generating one ATP. Propanol and propionate can be used as substrates for growth.

EUT BMCs

Ethanolamine utilization (EUT) BMCs are encoded in many diverse types of bacteria. Ethanolamine is cleaved to ammonia and acetaldehyde through the action of ethanolamine-ammonia lyase, which also requires vitamin B12 as a cofactor. Acetaldehyde is fairly volatile, and mutants deficient in the BMC shell have been observed to have a growth defect and release excess amounts of acetaldehyde. It has been proposed that sequestration of acetaldehyde in the metabolosome prevents its loss by volatility. The end-products of the EUT BMC are ethanol and acetyl-phosphate. Ethanol is likely a lost carbon source, but acetyl-phosphate can either generate ATP or be recycled to acetyl-CoA and enter the TCA cycle or several biosynthetic pathways.

Bifunctional PDU/EUT BMCs

Some bacteria, especially those in the genus *Listeria*, encode a single locus in which genes for both PDU and EUT BMCs are present. It is not yet clear whether this is truly a chimeric BMC with a mixture of both sets of proteins, or if two separate BMCs are formed.

Glycyl Radical Enzyme-containing BMCs (GRM)

Several different BMC loci have been identified that contain glycyl radical enzymes, which obtain the catalytic radical from the cleavage of s-adenosylcobalamin. One GRM locus in *Clostridium phytofermentans* has been shown to be involved in the fermentation of fucose and rhamnose, which are initially degraded to 1,2-propanediol under anaerobic conditions. The glycyl radical enzyme is proposed to dehydrate propanediol to propionaldehyde, which is then processed in a manner identical to the canonical PDU BMC.

Planctomycetes and Verrucomicrobia BMCs (PVM)

Distinct lineages of Planctomycetes and Verrucomicrobia encode a BMC locus. The locus in *Planctomyces limnophilus* has been shown to be involved in the aerobic degradation of fucose and rhamnose. An aldolase is thought to generate lactaldehyde, which is then processed through the BMC, resulting in 1,2-propanediol and lactyl-phosphate.

Rhodococcus and Mycobacterium BMCs (RMM)

Two types of BMC loci have been observed in members of the *Rhodococcus* and *Mycobacterium* genera, although their actual function has not been established. However, based on the characterized function of one of the genes present in the locus and the predicted functions of the other genes, it was proposed that these loci could be involved in the degradation of amino-2-propanol. The aldehyde generated in this predicted pathway would be the extremely toxic compound methylglyoxal; its sequestration within the BMC could protect the cell.

BMCs of Unknown Function (BUF)

One type of BMC locus does not contain RuBisCO or any of the core metabolosome enzymes, and has been proposed to facilitate a third category of biochemical transformations (i.e. not carbon fixation or aldehyde oxidation). The presence of genes predicted to code for amidohydrolases and deaminases could indicate that this BMC is involved in the metabolism of nitrogenous compounds.

Assembly

Carboxysomes

The assembly pathway for beta-carboxysomes has been identified, and begins with the protein CcmM nucleating RuBisCO. CcmM has two domains: an N-terminal gamma-carbonic anhydrase domain followed by a domain consisting of three to five repeats of RuBisCO small-subunit-like sequences. The C-terminal domain aggregates RuBisCO, likely by substituting for the actual RuBisCO small subunits in the L8-S8 holoenzyme, effectively cross-linking the RuBisCO in the cell into one large aggregate, termed the procarboxysome. The N-terminal domain of CcmM physically interacts with the N-terminal domain of the CcmN protein, which, in turn, recruits the hexagonal shell protein subunits via an encapsulation peptide on its C-terminus. Carboxysomes are then spatially aligned in the cyanobacterial cell via interaction with the bacterial cytoskeleton, ensuring their equal distribution into daughter cells.

Alpha-carboxysome assembly may be different than that of beta-carboxysomes, as they have no proteins homologous to CcmN or CcmM and no encapsulation peptides. Empty carboxysomes have been observed in electron micrographs. Some micrographs indicate that their assembly occurs as a simultaneous coalescence of enzymes and shell proteins as opposed to the seemingly stepwise fashion observed for beta-carboxysomes. The formation of simple alpha-carboxysomes in heterologous systems has been shown to require just Rubisco large and small subunits, the internal anchoring protein CsoS2 and the major shell protein CsoS1A.

Metabolosomes

Metabolosome assembly is likely similar to that of the beta-carboxysome, via an initial aggregation of the proteins to be encapsulated. The core proteins of many metabolosomes aggregate when expressed alone. Moreover, many encapsulated proteins contain terminal extensions that are strikingly similar to the C-terminal peptide of CcmN that recruits shell proteins. These encapsulation peptides are short (about 18 residues) and are predicted to form amphipathic alpha-helices. Some of these helices have been shown to mediate the encapsulation of native enzymes into BMCs, as well as heterologous proteins (such as GFP).

Regulation (Genetic)

With the exception of carboxysomes, in all tested cases, BMCs are encoded in operons that are expressed only in the presence of their substrate.

PDU BMCs in *Salmonella enterica* are induced by the presence of propanediol or glycerol under anaerobic conditions, and only propanediol under aerobic conditions. This induction is mediated by the global regulator proteins Crp and ArcA (sensing cyclic AMP and anaerobic conditions respectively), and the regulatory protein PocR, which is the transcriptional activator for both the *pdu* and the *cob* loci (the operon necessary for the synthesis of vitamin B12, a required cofactor for propanediol dehydratase).

EUT BMCs in *Salmonella enterica* are induced via the regulatory protein EutR by the simultaneous presence of ethanolamine and vitamin B12, which can happen under aerobic or anaerobic conditions. *Salmonella enterica* can only produce endogenous vitamin B12 under anaerobic conditions, although it can import cyanobalamin and convert it to vitamin B12 under either aerobic or anaerobic conditions.

PVM BMCs in *Planctomyces limnophilus* are induced by the presence of fucose or rhamnose under aerobic conditions, but not by glucose. Similar results were obtained for the GRM BMC from *Clostridium phytofermentans*, for which both sugars induce the genes coding for the BMC as well as the ones coding for fucose and rhamnose dissimilatory enzymes.

In addition to characterized regulatory systems, bioinformatics surveys have indicated that there are potentially many other regulatory mechanisms, even within a functional type of BMC (e.g. PDU), including two-component regulatory systems.

Relevance to Global and Human Health

Carboxysomes are present in all cyanobacteria and many other photo- and chemoautotrophic bacteria. Cyanobacteria are globally significant drivers of carbon fixation, and since they require carboxysomes to do so in current atmospheric conditions, the carboxysome is a major component of global carbon dioxide fixation.

Several types of BMCs have been implicated in virulence of pathogens, such as *Salmonella enterica* and *Listeria monocytogenes*. BMC genes tend to be upregulated under virulence conditions, and mutating them leads to a virulence defect as judged by competition experiments.

Biotechnological Applications

Several features of BMCs make them appealing for biotechnological applications. Because carboxysomes increase the efficiency of carbon fixation, much research effort has gone into introducing carboxysomes and required bicarbonate transporters into plant chloroplasts in order to engineer a chloroplastic CO_2 concentrating mechanism with some success.

More generally, because BMC shell proteins self-assemble, empty shells can be formed, prompting efforts to engineer them to contain customized cargo. Discovery of the encapsulation peptide on the termini of some BMC-associated proteins provides a means to begin to engineer custom BMCs by fusing foreign proteins to this peptide and co-expressing this with shell proteins. For example, by adding this peptide to pyruvate decarboxylase and alcohol dehydrogenase, researchers have engineered an ethanol bioreactor. Finally, the pores present in the shell proteins control the permeability of the shell: these can be a target for bioengineering, as they can be modified to allow the crossing of selected substrates and products.

Endospore

Bacterial endospores are special tough, dormant and resistant spores produced by some Gram-positive bacteria of Firmicute family during unfavorable environmental conditions. Endospores are developed within the vegetative cells (hence the name, endo = inside). They help the bacteria to endure the unfavorable environmental conditions.

Another importance of endospores is that it can be easily dispersed by wind, water and through the gut of animals. Bacillus and Clostridium are the most studied endospore forming bacterial genera. Bacillus enters into endospore formation cycle when the carbon or nitrogen source is getting limited in the growing medium.

Characteristics of Endospores

The endospores are structurally, metabolically and functionally very different from bacterial vegetative cells. The main characteristics of bacterial endospores are giving below:

- Endospores are exceptionally resistant to stressful environmental conditions such as heat, ultraviolet radiation, gamma radiation, chemical disinfectants and desiccation.

- Most of the endospores are viable for many years, even for 10, 000 years or more.

- Due to this long viability and their adaptations to stress conditions, most of the endospores producing bacteria are notorious pathogens.

- Wiping with alcohol or hydrogen peroxide or boiling at 100 C will not kill the bacterial endospores.

- However, endospores can be killed by autoclaving (at 121 C).

- Endospores can be visualized under light and electron microscope.

- Endospores will NOT take the usual bacterial stains such a safranin used in Gramstaining.

- Specific stains and special staining techniques are required to stain the endospores.

- The classically used stain to visualize endospore is Malachite Green and the staining procedure is known as Schaeffer–Fulton Staining.

- The endospore-producing mother cell is called sporangium. Ø Sporangium shows distinct differences from other vegetative cells.

- These characteristics are used for the identification purpose in bacterial taxonomy.

- The position of the endospore within the spore mother cell also varies.

- Based on the position of spores, the sporangium/spores may be Central spore, Subterminal spore, Terminal spore or Terminal spore with swollen sporangium.

- Both aerobic and anaerobic bacteria (of Gram-positive type) can produce endospores.

- No Archaebacteria are known to produce endospores.

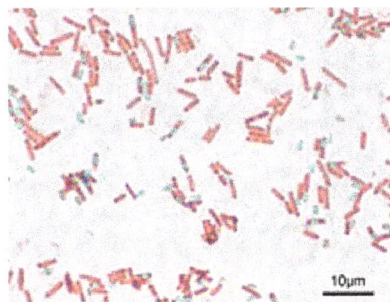

Structure of Endospores

- The structure of endospore is very complex since they possess multilayered coverings.

- The outermost layer of the spore is called exosporium which is relatively thin and delicate.

- Beneath the exosporium is a Spore Coat composed of several layers of proteins.

- Spore coat is comparatively thick.

- The thickness of the spore coat is one reason for the high resistance of endospores towards heat, radiation and chemicals.

- Inner to the spore coat is the Cortex.

- Cortex is the thicker wall layer in the endospores.

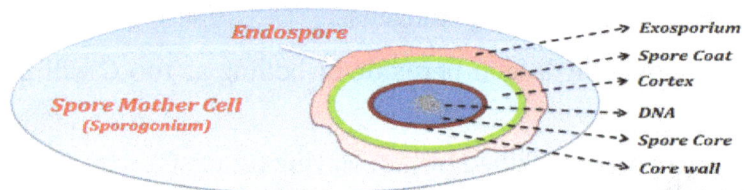

Bacterial Endospore- Diagrammatic

Bacterial Endospore-Diagrammatic

- Cortex is very large and sometimes occupy as much as half of the spore volume.

- The cortex is composed of peptidoglycan.

- The peptidoglycan in the cortex is less cross linked than that of vegetative cells.

- The innermost layer of the spore is called the Spore Cell wall or Core Cell Wall.

- Spore cell wall covers the central protoplast or Core of the endospore.

- The endospore core has a normal cell structure as that of a vegetative cell. Ø The core contains ribosomes and centrally placed nucleoid (genetic material).

- Unlike the vegetative cells, the core protoplast is metabolically inactive.

- The core only contains about 10 – 25% of water of the normal vegetative cell.

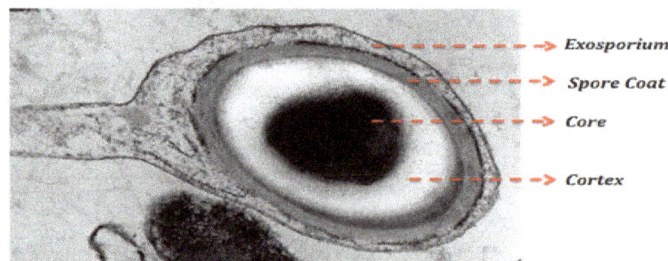

Bacterial Endospore

Bacterial Endospore

Bacterial Endospores are Extremely Resistant to Temperature, Radiations and Chemicals

The exact reason for the high resistance of endospores towards extreme temperature, radiation and chemicals is still unknown. Several explanations are now prevailing in the scientific community to explain this. Some of the possible explanations are given below:

- Endospores contain high amount of dipicolinic acid in its core (protoplast).

- In some endospores, about 15% of the total dry weight of the spore is contributed dipicolinic acid.

- The dipicolinic acid in bacterial endospore not occurs in free-state rather, it forms a complex with calcium ions (Ca).

- The calcium-dipicolinic acid can stabilize the genetic materials of the endospores.

- Large amounts of Small Acid Soluble DNA binding Proteins (SASPs) are reported to occur in the core of endospores.

- These proteins can bind to the DNA of endospores and can prevent the DNA from heat, radiations and chemicals. Ø The binding of SASPs to the DNA changes the molecular structure of DNA from its normal B-form to A-form.

- The A-DNA is more compact than B-DNA and thus A-DNA can have higher resistance against pyrimidine dimer formation by UV radiations.

- A-DNA is also comparatively more resistant to the denaturing effects of dry heat.

- The SASPs can also act as the carbon and nitrogen source of the newly formed vegetative cell during endospore germination.

- The cortex of the endospore can remove water from the core osmotically (cause dehydration). Ø Dehydration can provide heat resistance in bacterial cells.

- The thick spore coat can also act as an impermeable barrier against chemical such as hydrogen peroxides. Ø Spore coat can also restrict the entry of many hydrolyzing enzymes into the core.

- Bacterial endospores also contain a high amount of DNA repair enzymes.

- These repair enzymes can quickly heal all types of DNA lesions formed in the DNA when the spores are exposed to harsh environmental conditions.

Formation of Endospores in Bacteria

The process of formation of endospore is called Sporulation or Sporogenesis. Sporulation usually occurs when the bacterial cells face a nutrient deficient condition. The core of the endospore becomes increasingly dehydrated during the sporulation process. The formation of endospore is a complex process and it is completed in seven stages named as State – I (S-I) to Stage – VII (S-VII).

- S-I: Formation of axial filament: The genetic material of the bacterial cell is oriented in the exact center plane of the bacterial cell.

- **S-II: Formation of Septa:** A plasma membrane invagination grows into the lumen of the cell and forms a septum called forespore septum. The formation of septum results in the separation of a small portion of the DNA from rest of the genetic material.

- **S-III: Engulfment of the Forespore:** The membrane of the mother cell continues to grow and completely engulf the newly formed immature spore. Thus with the engulfment, the forespore is now covered by two plasma membrane and an inter membrane space.

- **S-IV: Formation of Cortex:** Cortex formation is started between the inter membrane space of the two membranes. Large amount of calcium and dipicolinic acid is also accumulated in Stage IV.

- **S-V: Formation of Protein Coat:** Protein coat is laid down over the cortex of the newly formed spore.

- **S-VI: Spore Maturation:** The core becomes increasingly dehydrated, the cell become metabolically inactive.

- **S-VII:** Enzymatic destruction of the sporogonium (spore mother cell) and release of the endospores.

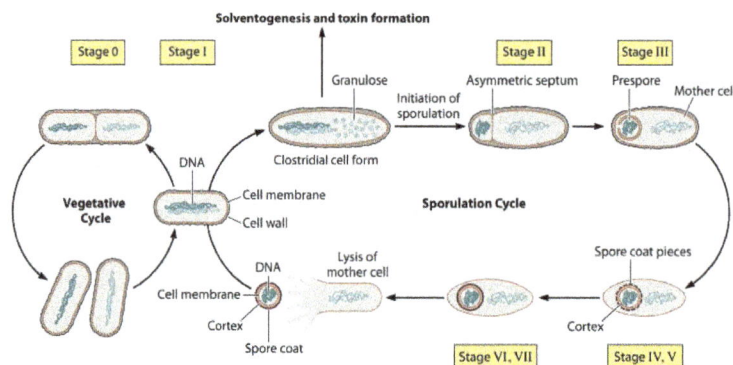

Germination of Endospores

The spore germination literally means the transformation of the dormant endospore to a metabolically active vegetative cell. The germination of spores occurs when the environmental conditions are suitable. Similar to the process of sporulation, the endospore germination is also a very complex event.

- The process of germination of endospore is completed in three stages:

  1). Activation

  2). Germination

  3). Outgrowth

1). Activation

- Activation of the endospore is a pre-request for its germination.

- Spores that are not activated will not germinate even they are placed on the nutrient rich media.

- Activation prepares the endospore for its germination (second step).

- Endospore activation is a reversible process.

- If the environmental conditions are not favorable, the activated spore can go back to its quiescent inactive stage.

- Activation of the endospores can be artificially induced by heat shock.

2). Germination

- Germination is the breaking of spore's dormant stage.

- Spore germination is characterized by the following events:

  o Swelling of the spore $ Rupture of the spore coat

  o Loss of resistance to heat or radiation

  o Loss of refactility

  o The release of spore components

  o The quick increase in the metabolic acidity of spores

- Endospore germination is an irreversible process.

- If the spore sense unfavorable conditions after it germination, it cannot be returned to its quiescent stage, rather it perishes.

- Germination of spores can be triggered by exposing the activated endospores to nutrients such as sugars or amino acids.

3). Out-growth

- It is the third stage of endospore germination.

- Spore completely emerges out from the spore coat.

- The protoplast of the spore completely exposed to the outer surroundings.

- They develop into an active vegetative cell.

Examples of Bacteria Producing Endospores

Aerobic endospore producing bacteria: Bacillus subtilis, B. megaterium, B. anthracis (cause Anthraxes).

Anaerobic endospore producing bacteria: Clostridium perfringes, C. tetani (cause tetanus) C. botulinum (cause botulism).

References

- Yeates, Todd O.; Kerfeld, Cheryl A.; Heinhorst, Sabine; Cannon, Gordon C.; Shively, Jessup M. (2008). "Protein-based organelles in bacteria: carboxysomes and related microcompartments". Nature Reviews Microbiology. 6 (9): 681–691. doi:10.1038/nrmicro1913. ISSN 1740-1526. PMID 18679172

- Bacterial-capsule-structure-and-importance-and-examples-of-capsulated-bacteria: microbeonline.com, Retrieved 12 May 2018

- "The Dublin City Rounders Alt-Country Song Contest Launched". IMRO.ie. Irish Music Rights Organisation. Retrieved 20 November 2016

- Cytoplasm-of-bacteria-6-components-microbiology-65026: biologydiscussion.com, Retrieved 28 March 2018

- P. Chen, D. I. Andersson & J. R. Roth (September 1994). "The control region of the pdu/cob regulon in Salmonella typhimurium". Journal of Bacteriology. 176(17): 5474–5482. PMC 196736. PMID 8071226

- Differences-between-fimbriae-and-pili: microbiologynotes.com, Retrieved 11 July 2018

- Kerfeld, Cheryl A.; Erbilgin, Onur (2015). "Bacterial microcompartments and the modular construction of microbial metabolism". Trends in Microbiology. 23 (1): 22–34. doi:10.1016/j.tim.2014.10.003. ISSN 0966-842X. PMID 25455419

- Bacterial-microcompartment, biochemistry-genetics-and-molecular-biology: sciencedirect.com, Retrieved 18 April 2018

- Axen, Seth D.; Erbilgin, Onur; Kerfeld, Cheryl A. (2014). "A Taxonomy of Bacterial Microcompartment Loci Constructed by a Novel Scoring Method". PLoS Computational Biology. 10 (10): e1003898. doi:10.1371/journal.pcbi.1003898. ISSN 1553-7358. PMC 4207490. PMID 25340524

Chapter 3

Bacterial Metabolism

Bacteria exhibit a wide range of metabolic types. These maybe based on the source of energy, source of carbon used for their growth or the electron donor used. An elaborate study of the fundamentals of bacterial metabolism has been provided in this chapter, which includes topics related to bacterial protein, anabolism, catabolism, oxidative phosphorylation etc.

Metabolism

Metabolism refers to all the biochemical reactions that occur in a cell or organism. The study of bacterial metabolism focuses on the chemical diversity of substrate oxidations and dissimilation reactions (reactions by which substrate molecules are broken down), which normally function in bacteria to generate energy. Also within the scope of bacterial metabolism is the study of the uptake and utilization of the inorganic or organic compounds required for growth and maintenance of a cellular steady state (assimilation reactions). These respective exergonic (energy-yielding) and endergonic (energy-requiring) reactions are catalyzed within the living bacterial cell by integrated enzyme systems, the end result being self-replication of the cell. The capability of microbial cells to live, function, and replicate in an appropriate chemical milieu (such as a bacterial culture medium) and the chemical changes that result during this transformation constitute the scope of bacterial metabolism.

The bacterial cell is a highly specialized energy transformer. Chemical energy generated by substrate oxidations is conserved by formation of high-energy compounds such as adenosine diphosphate (ADP) and adenosine triphosphate (ATP) or compounds containing the thioester bond,

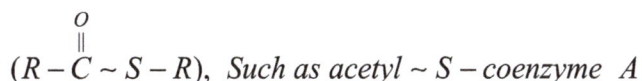

$$(R - \overset{\overset{O}{\|}}{C} \sim S - R), \; Such \; as \; acetyl \sim S - coenzyme \; A$$

(acetyl ~ SCoA) or succinyl ~ SCoA. ADP and ATP represent adenosine monophosphate (AMP) plus one and two high-energy phosphates (AMP ~ P and AMP ~ P~ P, respectively); the energy is stored in these compounds as high-energy phosphate bonds. In the presence of proper enzyme systems, these compounds can be used as energy sources to synthesize the new complex organic compounds needed by the cell. All living cells must maintain steady-state biochemical reactions for the formation and use of such high-energy compounds.

Heterotrophic Metabolism

Heterotrophic bacteria, which include all pathogens, obtain energy from oxidation of organic compounds. Carbohydrates (particularly glucose), lipids, and protein are the most commonly oxidized compounds. Biologic oxidation of these organic compounds by bacteria results in synthesis of ATP as the chemical energy source. This process also permits generation of simpler organic compounds (precursor molecules) needed by the bacteria cell for biosynthetic or assimilatory reactions.

The Krebs cycle intermediate compounds serve as precursor molecules (building blocks) for the energy-requiring biosynthesis of complex organic compounds in bacteria. Degradation reactions that simultaneously produce energy and generate precursor molecules for the biosynthesis of new cellular constituents are called amphibolic.

All heterotrophic bacteria require preformed organic compounds. These carbon- and nitrogen-containing compounds are growth substrates, which are used aerobically or anaerobically to generate reducing equivalents (e.g., reduced nicotinamide adenine dinucleotide; NADH + H$^+$); these reducing equivalents in turn are chemical energy sources for all biologic oxidative and fermentative systems. Heterotrophs are the most commonly studied bacteria; they grow readily in media containing carbohydrates, proteins, or other complex nutrients such as blood. Also, growth media may be enriched by the addition of other naturally occurring compounds such as milk (to study lactic acid bacteria) or hydrocarbons (to study hydrocarbon-oxidizing organisms).

Respiration

Glucose is the most common substrate used for studying heterotrophic metabolism. Most aerobic organisms oxidize glucose completely by the following reaction equation:

$$C_6H_{12}O_6 + 6O_2 \rightarrow 6CO_2 + 6H_2O + energy$$

This equation expresses the cellular oxidation process called respiration. Respiration occurs within the cells of plants and animals, normally generating 38 ATP molecules (as energy) from the oxidation of 1 molecule of glucose. This yields approximately 380,000 calories (cal) per mode of glucose (ATP ~ 10,000 cal/mole). Thermodynamically, the complete oxidation of one mole of glucose should yield approximately 688,000 cal; the energy that is not conserved biologically as chemical energy (or ATP formation) is liberated as heat (308,000 cal). Thus, the cellular respiratory process is at best about 55% efficient.

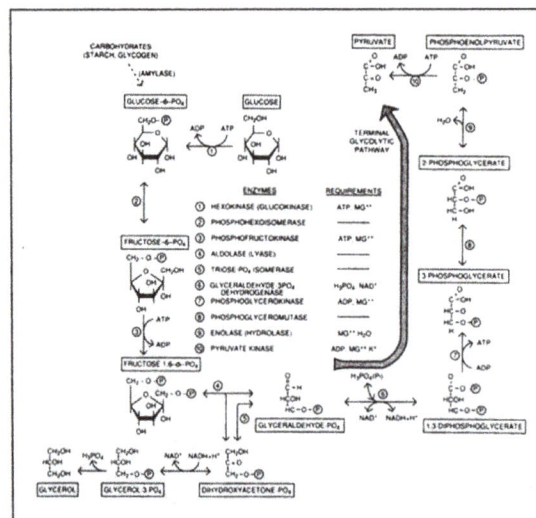

Figure: Glycolytic (EMP) pathway.

Glucose oxidation is the most commonly studied dissimilatory reaction leading to energy production or ATP synthesis. The complete oxidation of glucose may involve three fundamental biochemical pathways. The first is the glycolytic or Embden- Meyerhof-Parnas pathway, the second is the Krebs cycle (also called the citric acid cycle or tricarboxylic acid cycle), and the third is the series of membrane-bound electron transport oxidations coupled to oxidative phosphorylation.

Respiration takes place when any organic compound (usually carbohydrate) is oxidized completely to CO_2 and H_2O. In aerobic respiration, molecular O_2 serves as the terminal acceptor of electrons. For anaerobic respiration, NO_3-, SO_42-, CO_2, or fumarate can serve as terminal electron acceptors (rather than O_2), depending on the bacterium studied. The end result of the respiratory process is the complete oxidation of the organic substrate molecule, and the end products formed are primarily CO_2 and H_2O. Ammonia is formed also if protein (or amino acid) is the substrate oxidized. The biochemical pathways normally involved in oxidation of various naturally occurring organic compounds are summarized in figure below.

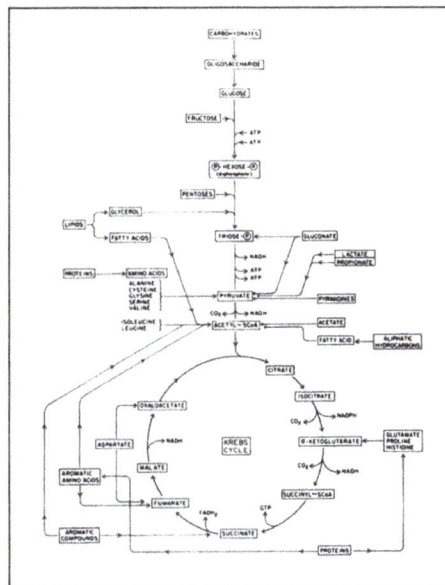

Figure: Heterotrophic metabolism, general pathway.

Metabolically, bacteria are unlike cyanobacteria (blue-green algae) and eukaryotes in that glucose oxidation may occur by more than one pathway. In bacteria, glycolysis represents one of several pathways by which bacteria can catabolically attack glucose. The glycolytic pathway is most commonly associated with anaerobic or fermentative metabolism in bacteria and yeasts. In bacteria, other minor heterofermentative pathways, such as the phosphoketolase pathway, also exist.

In addition, two other glucose-catabolizing pathways are found in bacteria: the oxidative pentose phosphate pathway (hexose monophosphate shunt), and the Entner-Doudoroff pathway, which is almost exclusively found in obligate aerobic bacteria. The highly oxidative Azotobacter and most Pseudomonas species, for example, utilize the Entner-Doudoroff pathway for glucose catabolism, because these organisms lack the enzyme phosphofructokinase and hence cannot synthesize fructose 1,6-diphosphate, a key intermediate compound in the glycolytic pathway. (Phospho-fructokinase is also sensitive to molecular O_2 and does not function in obligate aerobes). Other bacteria, which lack aldolase (which splits fructose-1,6-diphosphate into two triose phosphate compounds), also cannot have a functional glycolytic pathway. Although the Entner-Doudoroff pathway is usually associated with obligate aerobic bacteria, it is present in the facultative anaerobe Zymomonas mobilis (formerly Pseudomonas lindneri). This organism dissimilates glucose to ethanol and represents a major alcoholic fermentation reaction in a bacterium.

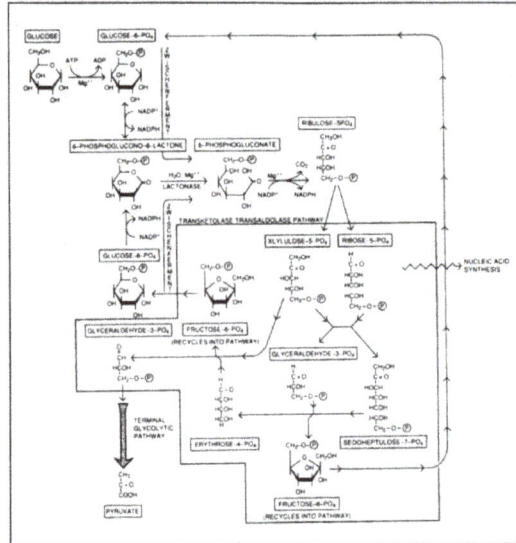

Figure: Hexose monophosphate (HMS) pathway.

Glucose dissimilation also occurs by the hexose monophosphate shunt. This oxidative pathway was discovered in tissues that actively metabolize glucose in the presence of two glycolytic pathway inhibitors (iodoacetate and fluoride). Neither inhibitor had an effect on glucose dissimilation, and NADPH + H$^+$ generation occurred directly from the oxidation of glucose-6-phosphate (to 6-phosphoglucono-δ-lactone) by glucose-6phosphate dehydrogenase. The pentose phosphate pathway subsequently permits the direct oxidative decarboxylation of glucose to pentoses. The capability of this oxidative metabolic system to bypass glycolysis explains the term shunt.

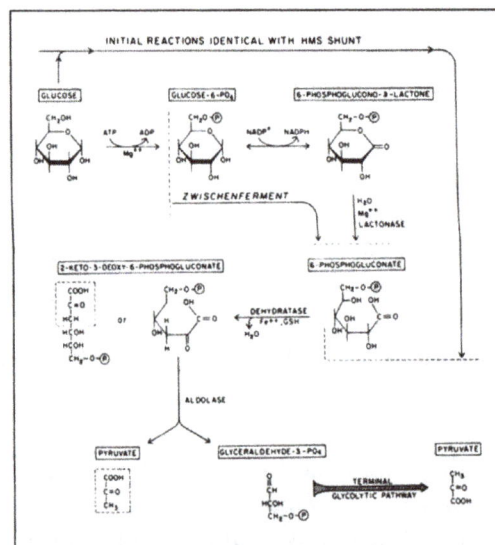

Figure: Entner-Doudoroff (ED) pathway.

The biochemical reactions of the Entner-Doudoroff pathway are a modification of the hexose mo-nophosphate shunt, except that pentose sugars are not directly formed. The two pathways are identical up to the formation of 6-phosphogluconate and then diverge. In the Entner-Doudoroff pathway, no oxidative decarboxylation of 6-phosphogluconate occurs and no pentose compound is formed. For this pathway, a new 6 carbon compound intermediate (2-keto-3-deoxy6-phospho-gluconate) is generated by the action of 6- phosphogluconate dehydratase (an $Fe^{2+}-$ and gluta-thione-stimulated enzyme); this intermediate compound is then directly cleaved into the triose (pyruvate) and a triose-phosphate compound (glyceraldehyde-3-phosphate) by the 2-keto-3-de-oxy6-phosphogluconate aldolase. The glyceraldehyde-3-phosphate is further oxidized to another pyruvate molecule by the same enzyme systems that catalyze the terminal glycolytic pathway.

The glycolytic pathway may be the major one existing concomitantly with the minor oxidative pen-tose phosphate - hexose monophosphate shunt pathway; the Entner-Doudoroff pathway also may function as a major pathway with a minor hexose monophosphate shunt. A few bacteria possess only one pathway. All cyanobacteria, *Acetobacter suboxydans*, and *A. xylinum* possess only the hexose monophosphate shunt pathway; *Pseudomonas saccharophilia* and *Z. mobilis* possess solely the Entner-Doudoroff pathway. Thus, the end products of glucose dissimilatory pathways are as follows:

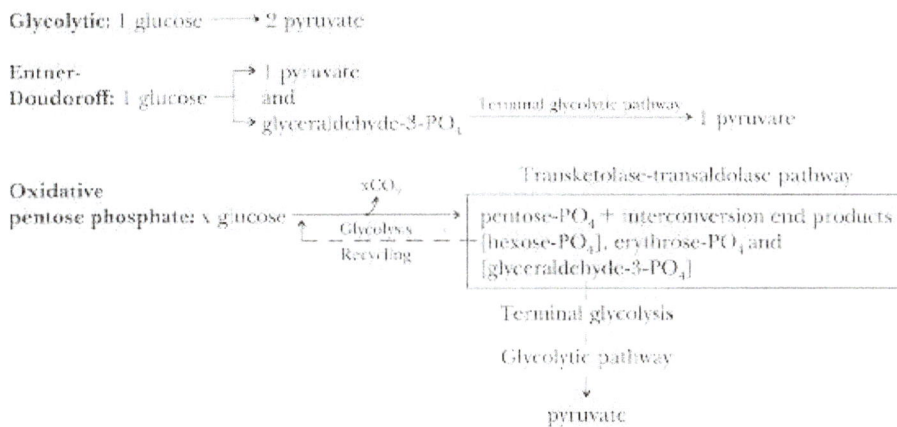

All major pathways of glucose or hexose catabolism have several metabolic features in common. First, there are the preparatory steps by which key intermediate compounds such as the tri-ose-PO_4, glyceraldehyde-3-phosphate, and/or pyruvate are generated. The latter two compounds are almost universally required for further assimilatory or dissimilatory reactions within the cell. Second, the major source of phosphate for all reactions involving phosphorylation of glucose or other hexoses is ATP, not inorganic phosphate (Pi). Actually, chemical energy contained in ATP must be initially spent in the first step of glucose metabolism (via kinase-type enzymes) to generate glucose-6-phosphate, which initiates the reactions involving hexose catabolism. Third, NADH + H^+ or NADPH + H^+ is generated as reducing equivalents (potential energy) directly by one or more of the enzymatic reactions involved in each of these pathways.

Fermentation

Fermentation, another example of heterotrophic metabolism, requires an organic compound as a terminal electron (or hydrogen) acceptor. In fermentations, simple organic end products are formed from the anaerobic dissimilation of glucose (or some other compound). Energy (ATP) is generated through the dehydrogenation reactions that occur as glucose is broken down enzymatically. The simple organic end products formed from this incomplete biologic oxidation process also serve as final electron and hydrogen acceptors. On reduction, these organic end products are secreted into the medium as waste metabolites (usually alcohol or acid). The organic substrate compounds are incompletely oxidized by bacteria, yet yield sufficient energy for microbial growth. Glucose is the most common hexose used to study fermentation reactions.

In the late 1850s, Pasteur demonstrated that fermentation is a vital process associated with the growth of specific microorganisms, and that each type of fermentation can be defined by the principal organic end product formed (lactic acid, ethanol, acetic acid, or butyric acid). His studies on butyric acid fermentation led directly to the discovery of anaerobic microorganisms. Pasteur concluded that oxygen inhibited the microorganisms responsible for butyric acid fermentation because both bacterial mobility and butyric acid formation ceased when air was bubbled into the fermentation mixture. Pasteur also introduced the terms aerobic and anaerobic.

In the experiments which we have described, fermentation by yeast is seen to be the direct consequence of the processes of nutrition, assimilation and life, when these are carried on without the agency of free oxygen. The heat required in the accomplishment of that work must necessarily have been borrowed from the decomposition of the fermentation matter. Fermentation by yeast appears, therefore, to be essentially connected with the property possessed by this minute cellular plant of performing its respiratory functions, somehow or other, with the oxygen existing combined in sugar.

For most microbial fermentations, glucose dissimilation occurs through the glycolytic pathway. The simple organic compound most commonly generated is pyruvate, or a compound derived enzymatically from pyruvate, such as acetaldehyde, α-acetolactate, acetyl ~ SCoA, or lactyl ~ SCoA. Acetaldehyde can then be reduced by NADH + H$^+$ to ethanol, which is excreted by the cell. The end product of lactic acid fermentation, which occurs in streptococci (e.g., Streptococcus lactis) and many lactobacilli (e.g., Lactobacillus casei, L. pentosus), is a single organic acid, lactic acid. Organisms that produce only lactic acid from glucose fermentation are homofermenters. Homofermentative lactic acid bacteria dissimilate glucose exclusively through the glycolytic pathway. Organisms that ferment glucose to multiple end products, such as acetic acid, ethanol, formic acid, and CO_2, are referred to as heterofermenters. Examples of heterofermentative bacteria include Lactobacillus, Leuconostoc, and Microbacterium species. Heterofermentative fermentations are more common among bacteria, as in the mixed-acid fermentations carried out by bacteria of the family Enterobacteriaceae (e.g., Escherichia coli, Salmonella, Shigella, and Proteus species). Many of these glucose fermenters usually produce CO_2 and H_2 with different combinations of acid end products (formate, acetate, lactate, and succinate). Other bacteria such as Enterobacter aerogenes, Aeromonas, Serratia, Erwinia, and Bacillus species also form CO_2 and H_2 as well as other neutral end products (ethanol, acetylmethylcarbinol, and 2,3-butylene glycol). Many obligately anaerobic clostridia (e.g., Clostridium saccharobutyricum, C. thermosaccharolyticum) and Butyribacterium species ferment glucose with the production of butyrate, acetate, CO_2, and H_2, whereas other Clostridum species (C. acetobutylicum and C. butyricum) also form these fermentation end products plus others (butanol, acetone, isopropanol, formate, and ethanol). Similarly, the anaerobic propionic acid bacteria (Propionibacterium species) and the related Veillonella species ferment glucose to form CO_2, propionate, acetate, and succinate. In these bacteria, propionate is formed by the partial reversal of the Krebs cycle reactions and involves a CO_2 fixation by pyruvate (the Wood-Werkman reaction) that forms oxaloacetate (a four-carbon intermediate). Oxaloacetate is then reduced to malate, fumarate, and succinate, which is decarboxylated to propionate. Propionate is also formed by another three-carbon pathway in C. propionicum, Bacteroides ruminicola, and Peptostreptococcus species, involving a lactyl ~ SCoA intermediate. The obligately aerobic acetic acid bacteria (Acetobacter and the related Gluconobacter species) can also ferment glucose, producing acetate and gluconate. Figure below summarizes the pathways by which the various major fermentation end products form from the dissimilation of glucose through the common intermediate pyruvate.

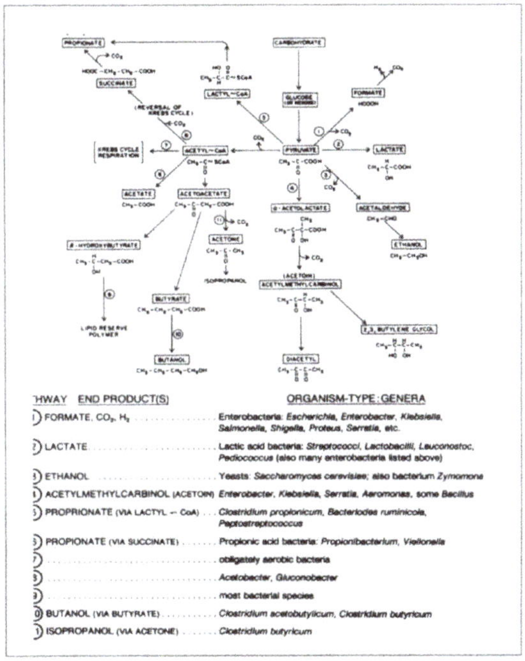

Figure: Fermentative pathways of bacteria and the major end products
formed with the organism type carrying out the fermentation.

For thermodynamic reasons, bacteria that rely on fermentative process for growth cannot generate as much energy as respiring cells. In respiration, 38 ATP molecules (or approximately 380,000 cal/mole) can be generated as biologically useful energy from the complete oxidation of 1 molecule of glucose (assuming 1 NAD(P)H = 3 ATP and 1 ATP $\rightarrow$ ADP + Pi = 10,000 cal/mole). Table below shows comparable bioenergetic parameters for the lactate and ethanolic fermentations by the glycolytic pathway. Although only 2 ATP molecules are generated by this glycolytic pathway, this is apparently enough energy to permit anaerobic growth of lactic acid bacteria and the ethanolic fermenting yeast, Saccharomyces cerevisiae. The ATP-synthesizing reactions in the glycolytic pathway specifically involve the substrate phosphorylation reactions catalyzed by phosphoglycerokinase and pyruvic kinase. Although all the ATP molecules available for fermentative growth are believed to be generated by these substrate phosphorylation reactions, some energy equivalents are also generated by proton extrusion reactions (acid liberation), which occur with intact membrane systems and involve the proton extrusion reactions of energy conservation as it applies to fermentative metabolism.

Table: Energy Obtained from Bacterial Fermentations by Substrate Phosphorylations.

	Fermentation	Actual Energy (Cal/mole)	Theoretical Energy (Cal/mole)	Efficiency (%)
Homolactic $Glycolysis$ $C_6H_{12}O_3 \rightarrow$ (*Lactic acid*)	$2CH_3 - \overset{H}{\underset{OH}{C}} - COOH$	$= 20,000$	57,000	35
Alconolic $Glycolysis$ $C_6H_{12}O_3 \rightarrow$ (*Glucose*)	$2CH_3 - \overset{H}{\underset{H}{C}} - CH + 2CO_3 +$ (*Ethanol*)	$= 20,000$	58,000	34

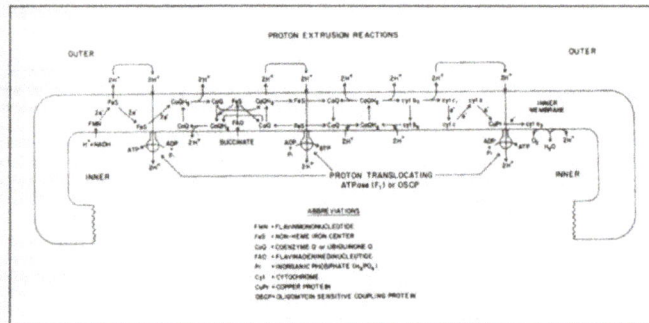

Figure: Mitchell hypotheses, a chemiosmotic model of energy transduction.

Krebs Cycle

The Krebs cycle (also called the tricarboxylic acid cycle or citic acid cycle) functions oxidatively in respiration and is the metabolic process by which pyruvate or acetyl ~ SCoA is completely decarboxylated to CO_2. In bacteria, this reaction occurs through acetyl ~ SCoA, which is the first product in the oxidative decarboxylation of pyruvate by pyruvate dehydrogenase. Bioenergetically, the following overall exergonic reaction occurs:

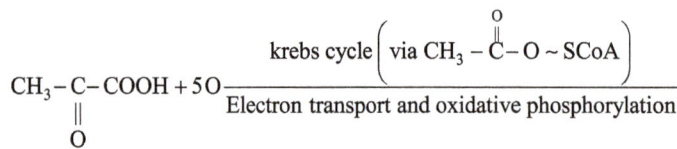

$$CH_3-\underset{\overset{\|}{O}}{C}-COOH + 5O \xrightarrow[\text{Electron transport and oxidative phosphorylation}]{\text{krebs cycle}\left(\text{via } CH_3-\overset{\overset{O}{\|}}{C}-O \sim SCoA\right)}$$

$$3CO_2 + 2H_2O + 15ATP\left(\cong 150,000\,cal\,/\,mole\right)$$

If 2 pyruvate molecules are obtained from the dissimilation of 1 glucose molecule, then 30 ATP molecules are generated in total. The decarboxylation of pyruvate, isocitrate, and α-ketoglutarate accounts for all CO_2 molecules generated during the respiratory process. Figure below shows the enzymatic reactions in the Krebs cycle. The chemical energy conserved by the Krebs cycle is contained in the reduced compounds generated (NADH + H[+], NADPH + H[+], and succinate). The potential energy inherent in these reduced compounds is not available as ATP until the final step of respiration (electron transport and oxidative phosphorylation) occurs.

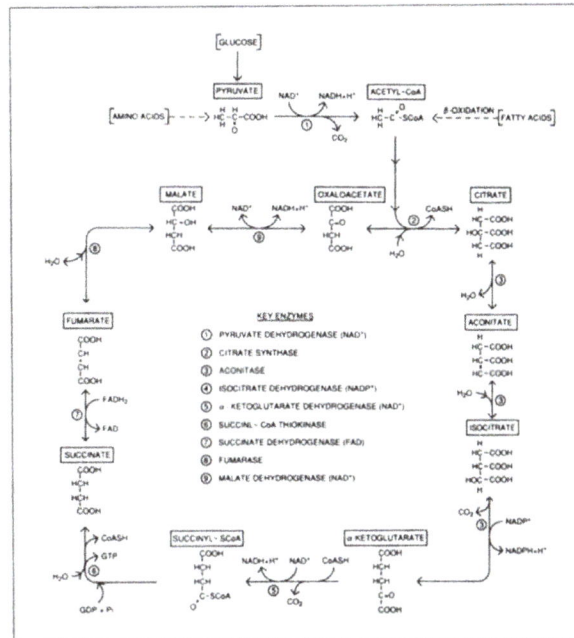

Figure: Krebs cycle (also tricarboxylic acid or citric acid cycle).

The Krebs cycle is therefore another preparatory stage in the respiratory process. If 1 molecule of pyruvate is oxidized completely to 3 molecules of CO_2, generating 15 ATP molecules, the oxidation of 1 molecule of glucose will yield as many as 38 ATP molecules, provided glucose is dissimilated by glycolysis and the Krebs cycle (further assuming that the electron transport/oxidative phosphorylation reactions are bioenergetically identical to those of eukaryotic mitochondria).

Glyoxylate Cycle

In general, the Krebs cycle functions similarly in bacteria and eukaryotic systems, but major differences are found among bacteria. One difference is that in obligate aerobes, L-malate may be oxidized directly by molecular O_2 via an electron transport chain. In other bacteria, only some Krebs cycle intermediate reactions occur because α-ketoglutarate dehydrogenase is missing.

A modification of the Krebs cycle, commonly called the glyoxylate cycle, or shunt, which exists in some bacteria. This shunt functions similarly to the Krebs cycle but lacks many of the Krebs cycle enzyme reactions. The glyoxylate cycle is primarily an oxidative pathway in which acetyl~SCoA is generated from the oxidation, of acetate, which usually is derived from the oxidation of fatty acids. The oxidation of fatty acids to acetyl~SCoA is carried out by the β-oxidation pathway. Pyruvate oxidation is not directly involved in the glyoxylate shunt, yet this shunt yields sufficient succinate and malate, which are required for energy production. The glyoxylate cycle also generates other precursor compounds needed for biosynthesis. The glyoxylate cycle was discovered as an unusual metabolic pathway during an attempt to learn how lipid (or acetate) oxidation in bacteria and plant seeds could lead to the direct biosynthesis of carbohydrates. The glyoxylate cycle converts oxaloacetate either to pyruvate and CO_2 (catalyzed by pyruvate carboxylase) or to phosphoenolpyruvate and CO_2 (catalyzed by the inosine triphosphate-dependent phosphoenolpyruvate carboxylase kinase). Either triose compound can then be converted to glucose by reversal of the glycolytic pathway. The glyoxylate cycle is found in many bacteria, including Azotobacter vinelandii and particularly in organisms that grow well in media in which acetate and other Krebs cycle dicarboxylic acid intermediates are the sole carbon growth source. One primary function of the glyoxylate cycle is to replenish the tricarboxylic and dicarboxylic acid intermediates that are normally provided by the Krebs cycle. A pathway whose primary purpose is to replenish such intermediate compounds is called anaplerotic.

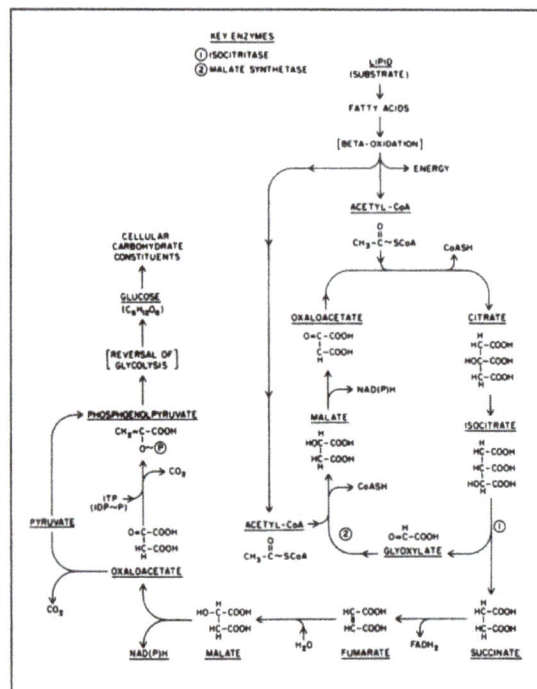

Figure: Glyoxylate shunt.

Electron Transport and Oxidative Phosphorylation

The final stage of respiration occurs through a series of oxidation-reduction electron transfer reactions that yield the energy to drive oxidative phosphorylation; this in turn produces ATP. The enzymes involved in electron transport and oxidative phosphorylation reside on the bacterial inner (cytoplasmic) membrane. This membrane is invaginated to form structures called respiratory vesicles, lamellar vesicles, or mesosomes, which function as the bacterial equivalent of the eukaryotic mitochondrial membrane.

Respiratory electron transport chains vary greatly among bacteria, and in some organisms are absent. The respiratory electron transport chain of eukaryotic mitochondria oxidizes NADH + H^+, NADPH + H^+, and succinate (as well as the coacylated fatty acids such as acetyl~SCoA). The bacterial electron transport chain also oxidizes these compounds, but it can also directly oxidize, via non-pyridine nucleotide-dependent pathways, a larger variety of reduced substrates such as lactate, malate, formate, α-glycerophosphate, H_2, and glutamate. The respiratory electron carriers in bacterial electron transport systems are more varied than in eukaryotes, and the chain is usually branched at the site(s) reacting with molecular O_2. Some electron carriers, such as nonheme iron centers and ubiquinone (coenzyme Q), are common to both the bacterial and mammalian respiratory electron transport chains. In some bacteria, the naphthoquinones or vitamin K may be found with ubiquinone. In still other bacteria, vitamin K serves in the absence of ubiquinone. In mitochondrial respiration, only one cytochrome oxidase component is found (cytochrome a + a_3 oxidase). In bacteria there are multiple cytochrome oxidases, including cytochromes a, d, o, and occasionally a + a_3.

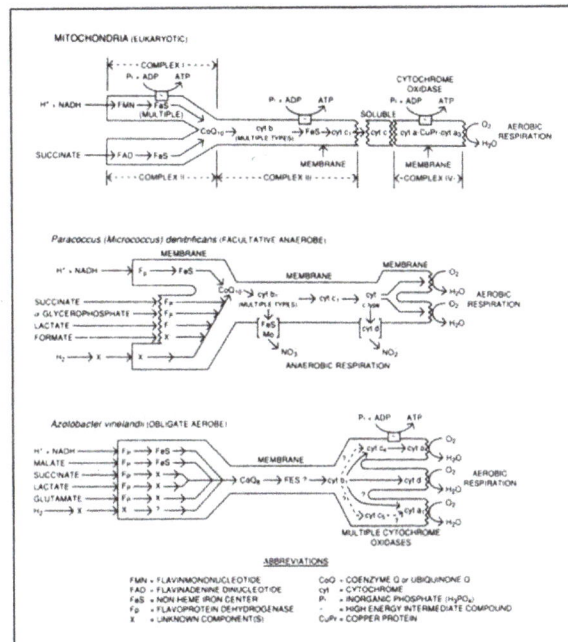

Figure: Respiratory electron transport chains.

In bacteria cytochrome oxidases usually occur as combinations of a1: d: o and a + a3:o. Bacteria also possess mixed-function oxidases such as cytochromes P-450 and P-420 and cytochromes c' and c'c', which also react with carbon monoxide. These diverse types of oxygen-reactive cytochromes undoubtedly have evolutionary significance. Bacteria were present before O_2 was formed; when O_2 became available as a metabolite, bacteria evolved to use it in different ways; this probably accounts for the diversity in bacterial oxygen-reactive hemoproteins.

Cytochrome oxidases in many pathogenic bacteria are studied by the bacterial oxidase reaction, which subdivides Gram-negative organisms into two major groups, oxidase positive and oxidase negative. This oxidase reaction is assayed for by using N,N,N', N'-tetramethyl-p-phenylenediamine oxidation (to Wurster's blue) or by using indophenol blue synthesis (with dimethyl-p-phenylenediamine and α-naphthol). Oxidase-positive bacteria contain integrated (cytochrome c type:oxidase) complexes, the oxidase component most frequently encountered is cytochrome o, and occasionally a + a_3. The cytochrome oxidase responsible for the indophenol oxidase reaction complex was isolated from membranes of Azotobacter vinelandii, a bacterium with the highest respiratory rate of any known cell. The cytochrome oxidase was found to be an integrated cytochrome c4:0 complex, which was shown to be present in Bacillus species. These Bacillus strains are also highly oxidase positive, and most are found in morphologic group II.

Both bacterial and mammalian electron transfer systems can carry out electron transfer (oxidation) reactions with NADH + H$^+$, NADPH + H$^+$, and succinate. Energy generated from such membrane oxidations is conserved within the membrane and then transferred in a coupled manner to drive the formation of ATP. The electron transfer sequence is accomplished entirely by membrane-bound enzyme systems. As the electrons are transferred by a specific sequence of electron carriers, ATP is synthesized from ADP + inorganic phosphate (Pi) or orthophosphoric acid (H$_3$PO$_4$).

In respiration, the electron transfer reaction is the primary mode of generating energy; electrons (2e-) from a low-redox-potential compound such as NADH + H$^+$ are sequentially transferred to a specific flavoprotein dehydrogenase or oxidoreductase (flavin mononucleotide type for NADH or flavin adenine dinucleotide type for succinate); this electron pair is then transferred to a nonheme iron center (FeS) and finally to a specific ubiquinone or a naphthoquinone derivative. This transfer of electrons causes a differential chemical redox potential change so that within the membrane enough chemical energy is conserved to be transferred by a coupling mechanism to a high-energy compound (e.g., ADP + Pi → ATP). ATP molecules represent the final stable high-energy intermediate compound formed.

A similar series of redox changes also occurs between ubiquinone and cytochrome c, but with a greater differential in the oxidation-reduction potential level, which allows for another ATP synthesis step. The final electron transfer reaction occurs at the cytochrome oxidase level between reduced cyotchrome c and molecular O$_2$; this reaction is the terminal ATP synthesis step.

Mitchell or Proton Extrusion Hypothesis

A highly complex but attractive theory to explain energy conservation in biologic systems is the chemiosmotic coupling of oxidative and photosynthetic phosphorylations, commonly called the Mitchell hypothesis. This theory attempts to explain the conservation of free energy in this process on the basis of an osmotic potential caused by a proton concentration differential (or proton gradient) across a proton-impermeable membrane. Energy is generated by a proton extrusion reaction during membrane-bound electron transport, which in essence serve as a proton pump; energy conservation and coupling follow. This represents an obligatory "intact" membrane phenomenon. The energy thus conserved (again within the confines of the membrane and is coupled to ATP synthesis. This would occur in all biologic cells, even in the lactic acid bacteria that lack a cytochrome-dependent electron transport chain but still possesses a cytoplasmic membrane. In this hypothesis, the membrane allows for charge separation, thus forming a proton gradient that drives all bioenergization reactions. By such means, electromotive forces can be generated by oxidation-reduction reactions that can be directly coupled to ion translocations, as in the separation of H$^+$ and OH$^-$ ions in electrochemical systems. Thus, an enzyme or an electron transfer carrier on a membrane that undergoes an oxidation-reduction reaction serves as a specific conductor for OH$^-$ (or O^{2-}), and "hydrodehydration" provides electromotive power, as it does in electrochemical cells.

The concept underlying Mitchell's hypothesis is complex, and many modifications have been proposed, but the theory's most attractive feature is that it unifies all bioenergetic conservation principles into a single concept requiring an intact membrane vesicle to function properly. Figure 4-9 shows how the Mitchell hypothesis might be used to explain energy generation, conservation, and transfer by a coupling process. The least satisfying aspect of the chemiosmotic hypothesis is the lack of understanding of how chemical energy is actually conserved within the membrane and how it is transmitted by coupling for ATP synthesis.

Bacterial Photosynthesis

Many prokaryotes (bacteria and cyanobacteria) possess phototrophic modes of metabolism. The types of photosynthesis in the two groups of prokaryotes differ mainly in the type of compound that serves as the hydrogen donor in the reduction of CO_2 to glucose. Phototrophic organisms differ from heterotrophic organisms in that they utilize the glucose synthesized intracellularly for biosynthetic purposes (as in starch synthesis) or for energy production, which usually occurs through cellular respiration.

Unlike phototrophs, heterotrophs require glucose (or some other preformed organic compound) that is directly supplied as a substrate from an exogenous source. Heterotrophs cannot synthesize large concentrations of glucose from CO_2 by specifically using H_2O or (H_2S) as a hydrogen source and sunlight as energy. Plant metabolism is a classic example of photolithotrophic metabolism: plants need CO_2 and sunlight; H_2O must be provided as a hydrogen source and usually NO_3- is the nitrogen source for protein synthesis. Organic nitrogen, supplied as fertilizer, is converted to NO_3- in all soils by bacteria via the process of ammonification and nitrification. Although plant cells are phototrophic, they also exhibit a heterotrophic mode of metabolism in that they respire. For example, plants use classic respiration to catabolize glucose that is generated photosynthetically. Mitochondria as well as the soluble enzymes of the glycolytic pathway are required for glucose dissimilation, and these enzymes are also found in all plant cells. The soluble Calvin cycle enzymes, which are required for glucose synthesis during photosynthesis, are also found in plant cells. It is not possible to feed a plant by pouring a glucose solution on it, but water supplied to a plant will be "photolysed" by chloroplasts in the presence of light; the hydrogens generated from H_2O is used by Photosystems I and II (PSI and PSII) to reduce $NADP^+$ to NADPH + H+. With the ATP generated by PSI and PSII, these reduced pyridine nucleotides, CO_2 is reduced intracellularly to glucose. This metabolic process is carried out in an integrated manner by Photosystems I and II ("Z" scheme) and by the Calvin cycle pathway. A new photosynthetic, and nitrogen fixing bacterium, Heliobacterium chlorum, staining Gram positive was isolated, characterized, and found to contain a new type of chlorophyll, i.e., bacteriochlorophyll 'g'. 16S r-RNA sequence analyses showed this organism to be phylogenetically related to members of the family Bacillaceae, although all currently known phototrophes are Gram negative. A few Heliobacteriium strains did show the presence of endospores. Another unusual phototrophe is the Gram negative Halobacterium halobium (now named Halobacterium salinarium), an archaebacterium growing best at 30°C in 4.0–5.0 M (or 25%, w/v) NaCl. This bacterium is a facultative phototrophe having a respiratory mode; it also possesses a purple membrane within which bacteriorhodopsin serves as the active photosynthetic pigment. This purple membranae possesses a light driven proton translocation pump which mediates photosynthetic ATP synthesis via a proton extrusion reaction. Table below summarizes the characteristics of known photosynthetic bacteria.

Autotrophy

Bacteria that grow solely at the expense of inorganic compounds (mineral ions), without using sunlight as an energy source, are called autotrophs, chemotrophs, chemoautotrophs, or chemolithotrophs. Like photosynthetic organisms, all autotrophs use CO_2 as a carbon source for growth; their nitrogen comes from inorganic compounds such as NH_3, NO_3-, or N_2. Interestingly, the energy source for such organisms is the oxidation of specific inorganic compounds. Which inorganic compound is oxidized depends on the bacteria in question. Many autotrophs will not grow on media that contain organic matter, even agar.

Also found among the autotrophic microorganisms are the sulfur-oxidizing or sulfur-compound-oxidizing bacteria, which seldom exhibit a strictly autotrophic mode of metabolism like the obligate nitrifying bacteria. The representative sulfur compounds oxidized by such bacteria are H_2S, S_2, and S_2O_3. Among the sulfur bacteria are two very interesting organisms; Thiobacillus ferrooxidans, which gets its energy for autotrophic growth by oxidizing elemental sulfur or ferrous iron, and T. denitrificans, which gets its energy by oxidizing S_2O_3 anaerobically, using NO_3- as the sole terminal electron acceptor. T denitrificans reduces NO_3 to molecular N_2, which is liberated as a gas; this biologic process is called denitrification.

All autotrophic bacteria must assimilate CO_2, which is reduced to glucose from which organic cellular matter is synthesized. The energy for this biosynthetic process is derived from the oxidation of inorganic compounds discussed in the previous paragraph. Note that all autotrophic and phototrophic bacteria possess essentially the same organic cellular constituents found in heterotrophic bacteria; from a nutritional viewpoint, however, the autotrophic mode of metabolism is unique, occurring only in bacteria.

Anerobic Respiration

Some bacteria exhibit a unique mode of respiration called anaerobic respiration. These heterotrophic bacteria that will not grow anaerobically unless a specific chemical component, which serves as a terminal electron acceptor, is added to the medium. Among these electron acceptors are NO_3-, SO_42-, the organic compound fumarate, and CO_2. Bacteria requiring one of these compounds for anaerobic growth are said to be anaerobic respirers.

A large group of anaerobic respirers are the nitrate reducers. The nitrate reducers are predominantly heterotrophic bacteria that possess a complex electron transport system(s) allowing the NO3− ion to serve anaerobically as a terminal acceptor of electrons.

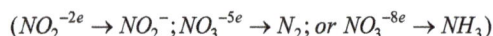

$$(NO_2^{-2e} \rightarrow NO_2^-; NO_3^{-5e} \rightarrow N_2; \text{ or } NO_3^{-8e} \rightarrow NH_3)$$

The organic compounds that serve as specific electron donors for these three known nitrate reduction processes are shown in table below. The nitrate reductase activity is common in bacteria and is routinely used in the simple nitrate reductase test to identify bacteria.

The organic compounds that serve as specific electron donors for these three known nitrate reduction processes are shown in table below. The nitrate reductase activity is common in bacteria and is routinely used in the simple nitrate reductase test to identify bacteria.

Table: Nitrate Reducers

Physiologic Types of Nitrate Reductases	Electron Donor(s)	Representative Organisms
Respiratory	Formate	Escherichia coli
$\left(NO_3^- \to NO_2^- \right)$	NADH	Klebsiella acrogenes
Denitrifying	NADH	Pseudomonas aeroginosa
	Pyruvate	Clostridium perfringens
$\left(NO_3^- \to NO_2 \right)$	NADH,succinate	paracoccus denitrificans
Assimilatory	Lactate	Staphylococcus aureus
$NO_3^- \to N$	H_{2+} formate	Vibrio succinogenes
	NADH,succinate	Enterobacter acrogens
	NADH	Escherichia coli
	NADH,lactate,glycerol-phosphate	

$$4\,AH_2 + HNO_3 \xrightarrow{\text{Nitrate reduction}} 4A + NH_3 + 3H_2O + \text{energy}$$

$(AH_2 = $ organic substrate, which serves as electron donor$)$

A second group of anaerobic respirers, the sulfate reducers, utilizes SO_4^{2-} ion in similar fashion

$(SO_4^{2-\,8e^-}\ H_2S)$:

$$4\,AH_2 + H_2SO_4 \xrightarrow{\text{Sulfate reduction}} 4A + H_2S + 4H_2O + \text{energy}$$

The third group, the fumarate respirers, are anaerobic bacteria that require exogenous, $HOOC- CH = CH - COOH$ for growth. Fumarate is reduced to succinate, $(HOOC- CH_2 - CH_2 - COOH)$, Which is secre ted as a by- product.

$\left(NO_3^- \to NO_3 \right)$

$3CO_2 + 2H_2O + 15\,ATP\left(\cong 150,000\,cal\,/\,mole \right)$

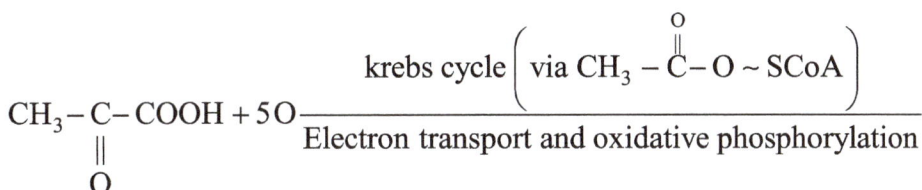

$$CH_3 - C - COOH + 5O \underbrace{\frac{\text{krebs cycle}\left(\text{via } CH_3 - \overset{\overset{\displaystyle O}{\|}}{C} - O \sim SCoA \right)}{\text{Electron transport and oxidative phosphorylation}}}_{}$$
$\quad\ \ \underset{O}{\overset{\|}{}}$

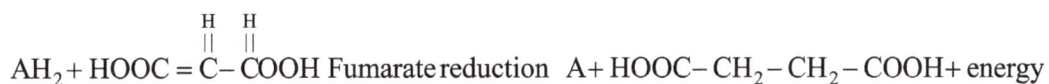

$$AH_2 + HOOC = \overset{\overset{\displaystyle H\ \ H}{\| \ \ \|}}{C} - COOH \ \underline{\text{Fumarate reduction}} \ A + HOOC- CH_2 - CH_2 - COOH + \text{energy}$$

Organisms of still another specialized group of anaerobic respirers, the methanogens, produce methane gas,

$CO_2 \xrightarrow{8e} CH_4$,

as a metabolic end product of microbial growth. H_2 gas is the growth substrate; CO_2 is the terminal electron acceptor.

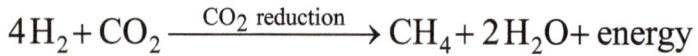

$$4H_2 + CO_2 \xrightarrow{CO_2 \text{ reduction}} CH_4 + 2H_2O + \text{energy}$$.

The methanogens are among the most anaerobic bacteria known, being very sensitive to small concentrations of molecular O_2. They are also archaebacteria, which typically live in unusual and deleterious environments.

All of the above anaerobic respirers obtain chemical energy for growth by using these anaerobic energy-yielding oxidation reactions.

The Nitrogen Cycle

Nowhere can the total metabolic potential of bacteria and their diverse chemical-transforming capabilities be more fully appreciated than in the geochemical cycling of the element nitrogen. All the basic chemical elements (S, O, P, C, and H) required to sustain living organisms have geochemical cycles similar to the nitrogen cycle.

The nitrogen cycle is an ideal demonstration of the ecologic interdependence of bacteria, plants, and animals. Nitrogen is recycled when organisms use one form of nitrogen for growth and excrete another nitrogenous compound as a waste product. This waste product is in turn utilized by another type of organism as a growth or energy substrate. Figure below shows the nitrogen cycle.

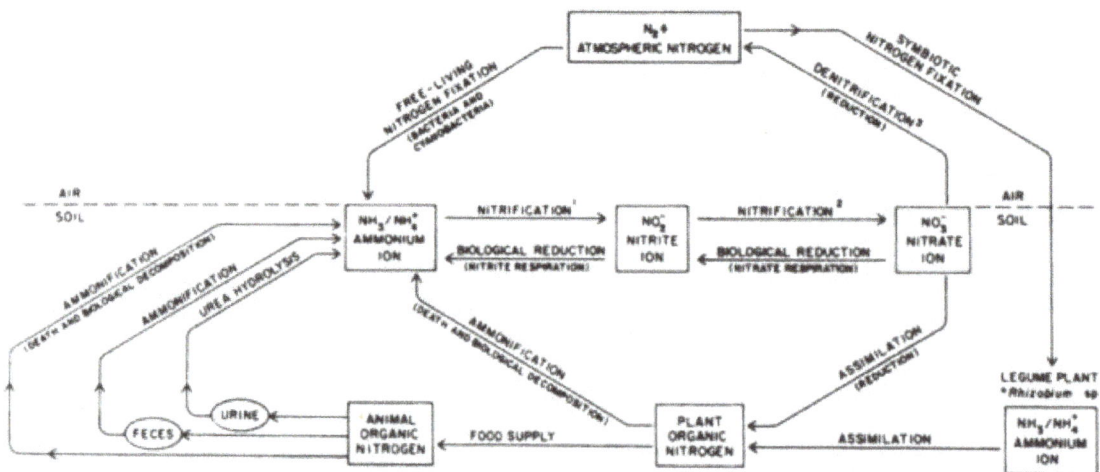

[1] Carried out primarily by chemoautotrophes of the genus Nitrosomonas

[2] Carried out by chemoautotrophes of the genus Nitrobacter

[3] Carried out by denitrifying bacteria, a property exhibited by some heterotrophic bacteria

Figure: The nitrogen cycle.

When the specific breakdown of organic nitrogenous compounds occurs, that is, when proteins are degraded to amino acids (proteolysis) and then to inorganic NH_3, by heterotrophic bacteria, the process is called ammonification. This is an essential step in the nitrogen cycle. At death, the organic constituents of the tissues and cells decompose biologically to inorganic constituents by a process called mineralization; these inorganic end products can then serve as nutrients for other life forms. The NH_3 liberated in turn serves as a utilizable nitrogen source for many other bacteria. The breakdown of feces and urine also occurs by ammonification.

The other important biologic processes in the nitrogen cycle include nitrification (the conversion of NH_3 to NO_3 by autotrophes in the soil; denitrification (the anaerobic conversion of NO_3 to N_2 gas) carried out by many heterotrophs); and nitrogen fixation (N_2 to NH_3, and cell protein). The latter is a very specialized prokaryotic process called diazotrophy, carried out by both free-living bacteria (such as Azotobacter, Derxia, Beijeringeia, and Azomona species) and symbionts (such as Rhizobium species) in conjunction with legume plants (such as soybeans, peas, clover, and bluebonnets). All plant life relies heavily on NO^{3-} as a nitrogen source, and most animal life relies on plant life for nutrients.

Catabolism

Bacterial catabolism comprises the biochemical activities concerned with the net breakdown of complex substances to simpler substances by living cells. Substances with a high energy level are converted to substances of low energy content, and the organism utilizes a portion of the released energy for cellular processes. Endogenous catabolism relates to the slow breakdown of nonvital intracellular constituents to secure energy and replacement building blocks for the maintenance of the structural and functional integrity of the cell. This ordinarily occurs in the absence of an external supply of food. Exogenous catabolism refers to the degradation of externally available food. The principal reactions employed are dehydrogenation or oxygenation (either represents biological oxidation), hydrolysis, hydration, decarboxylation, and intermolecular transfer and substitution. The complete catabolism of organic substances results in the formation of carbon dioxide, water, and other inorganic compounds and is known as mineralization. Catabolic processes may degrade a substance only part way. The resulting intermediate compounds may be reutilized in biosynthetic processes, or they may accumulate intra- or extracellularly. Catabolism also implies a conversion of the chemical energy into a relatively few energy-rich compounds or "bonds," in which form it is biologically useful; also, part of the chemical energy is lost as heat.

Bacterial intermediary metabolism relates to the chemical steps involved in metabolism between the starting substrates and the final product. Normally these intermediates, or precursors of subsequent products, do not accumulate inside or outside the bacterial cell in significant amounts, being transformed serially as rapidly as they are formed. The identification of such compounds, the establishment of the coenzymes and enzymes catalyzing the individual reaction steps, the identity of active forms of the intermediates, and other details of the reaction mechanisms are the objectives of a study of bacterial intermediary metabolism.

Many bacteria are able to decompose organic compounds and to grow in the absence of oxygen gas. Such anaerobic bacteria obtain energy and certain organic compounds needed for growth by a process of fermentation. This consists of an oxidation of a suitable organic compound, using another organic compound as an oxidizing agent in place of molecular oxygen. In most fermentations both the compounds oxidized and the compounds reduced (used as an oxidizing agent) are derived from a single fermentable substrate. In other fermentations, one substrate is oxidized and another is reduced. Different bacteria ferment different substrates. Many bacteria are able to ferment carbohydrates such as glucose and sucrose, polyalchohols such as mannitol, and salts of organic acids such as pyruvate and lactate. Other compounds, such as cellulose, amino acids, and purines, are fermented by some bacteria.

Protein and Amino Acid Catabolism

Some bacteria and fungi particularly pathogenic, food spoilage and soil microorganisms can use proteins as their source of carbon and energy.

1. Proteases are enzymes that break down proteins into amino acids.

2. Amino acids are deaminated, and then enter the Kreb's Cycle.

Intact proteins cannot cross bacterial plasma membrane, so bacteria must produce extracellular enzymes called proteases and peptidases that break down the proteins into amino acids, which can enter the cell. Many of the amino acids are used in building bacterial proteins, but some may also be broken down for energy. If this is the way amino acids are used, they are broken down to some form that can enter the Kreb's cycle. These reactions include:

1. Deamination or Transamination —the amino group is removed or transferred, or converted to an ammonium ion, and excreted. The remaining organic acid (the part of the amino acid molecule that is left after the amino group is removed) can enter the Kreb's cycle.

2. Decarboxylation —the ---COOH group is removed.

3. Dehydrogenation —a hydrogen is removed.

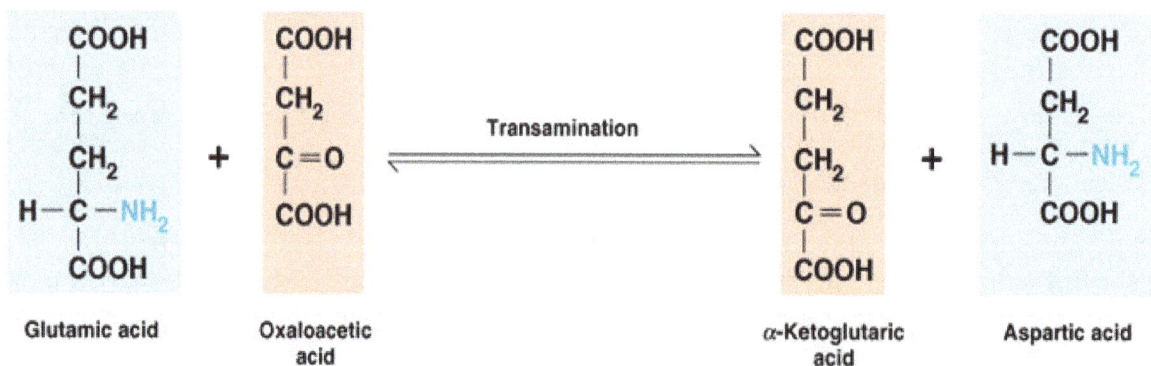

(b) Process of transamination

Figure: Process of Transamination

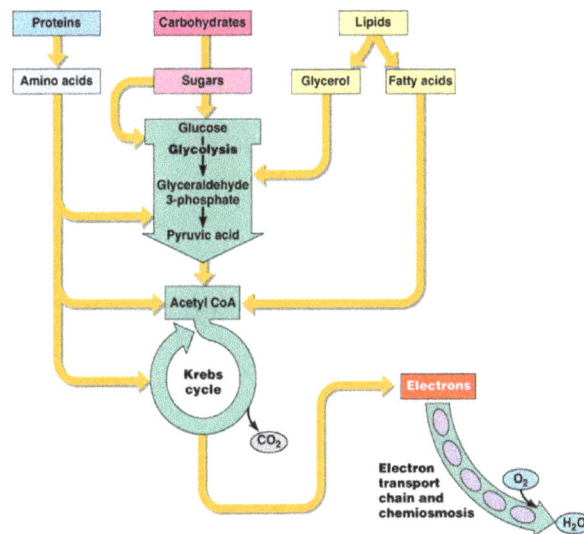

Figure: Overview of catabolism of Organic Acids.

Lipid Catabolism

Microorganisms frequently use lipids as energy sources. Triglycerides or triacylglycerols, esters of glycerol and fatty acids, are common energy sources. They can be hydrolyzed to glycerol and fatty acids by microbial lipases. The glycerol is then phosphorylated, oxidized to dihyroxyacetone phosphate, and catabolised in the glycolytic pathway.

1. Lipases are enzymes that break down fats into fatty acid and glycerol components

2. Beta oxidation is the breakdown of fatty acids into two carbon segments (acetyl CoA),

Which can enter the Krebs cycle.

Functions of Lipids in Microbes

- Lipids are essential to the structure and function of membranes

- Lipids also function as energy reserves, which can be mobilized as sources of carbon

- 90% of this lipid is "triacyglycerol"

  Triacyglycerol -----lipase-----> glycerol + 3 fatty acids

- The major fatty acid metabolism is "β-oxidation"

Lipids are broken down into their constituents of glycerol and fatty acids.

Glycerol is oxidised by glycolysis and the TCA cycle:

- Bacteria are capable of growth on fatty acids and lipids. Lipids are part of the membranes of living organisms and if available (usually because the organism that was using them dies) can be used as a food source.

- Lipids are large molecules and cannot be transported across the membrane.

- A class of extracellular enzymes called lipases are responsible for the breakdown of lipids. Lipases attack the bond between the glycerol molecule oxygen and the fatty acid.

- Phospholipids are attacked by phospholipases. There are four classes of phospholipases given different names depending upon the bond they cleave. Phospholipases are not particular about their substrate and will attack a glycerol ester linkage containing any length fatty acid attached to it. The result of this digestion is a hydrophillic head molecule, glycerol and fatty acids of various chain lengths.

- The head can be one of several small organic molecules that are funneled into the TCA cycle by one or two reactions that we won't cover here.

- Glycerol is converted into 3-Phosphoglycerate (depending upon the action of phospholipase C or phospholipase D) and eventually pyruvate via glycolysis.

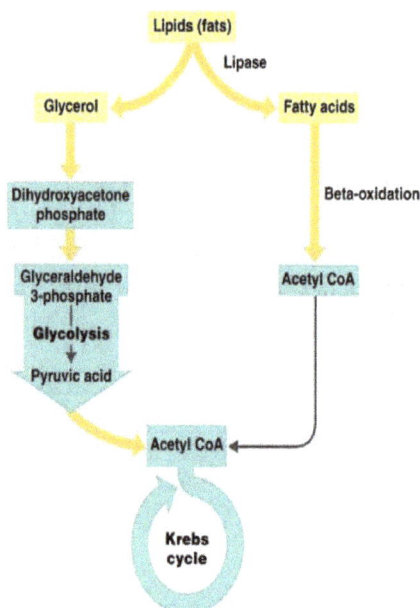

Figure: Lipid Catabolism

The β- oxidation Pathway

Characteristic features;

- Every other carbon is converted to a C=O

- Allows nucleophilic attack by CoA-SH on remaining chain

- 1 CoA is used for every 2 carbon segment to release acetyl-CoA

- Each round produces

1 FADH2 , 1 NADH, 1 Acetyl-CoA (2 in the Last Round)

Step 1: Dehydrogenation of Alkane to Alkene Catalyzed by isoforms of acyl-CoA dehydrogenase (AD) on the inner mitochondrial membrane.

Step 2: Hydration of Alkene Catalyzed by two isoforms of enoyl-CoA hydratase:

Soluble short-chain hydratase (crotonase) Membrane-bound long-chain hydratase, part of tri-functional complex Water adds across the double bond yielding alcohol.

Step 3: Dehydrogenation of Alcohol Catalyzed by β-hydroxyacyl-CoA dehydrogenase The enzyme uses NAD cofactor as the hydride acceptor Only L-isomers of hydroxyacyl CoA act as substrates Analogous to malate dehydrogenase reaction in the CAC.

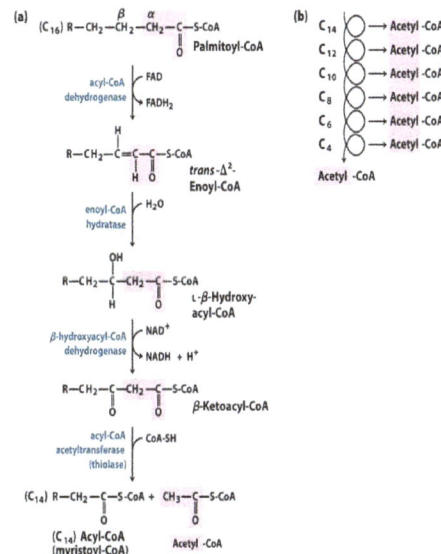

Figure: The β- oxidation pathway

Step 4: Transfer of Fatty Acid Chain Catalyzed by acyl-CoA acetyltransferase (thiolase) via covalent mechanism, The carbonyl carbon in b -ketoacyl-CoA is electrophilic Active site thiolate acts as nucleophile and releases acetyl-CoA; Terminal sulfur in CoA-SH acts as nucleophile.

The fatty acid is now two carbons shorter and an Acetyl-CoA, has been generated which can be fed into the TCA cycle. The smaller fatty acid moves through the β-oxidation pathway again, producing another Acetyl-CoA and shrinking by 2 carbons.

By performing successive rounds of beta oxidation on a fatty acid, it is possible to convert it completely to Acetyl-CoA. Often fatty acids with odd numbers of carbons, the final reaction will yield acetyl-CoA and Coenzyme-A hooked to a three carbon fatty acid (propionyl-CoA). Propionyl-CoA is handled differently by different bacteria. In E. coli it is converted into pyruvate.

Anabolism

Anabolism is the process by which the body utilizes the energy released by catabolism to synthesize complex molecules. These complex molecules are then utilized to form cellular structures that are formed from small and simple precursors that act as building blocks.

Stages of Anabolism

There are three basic stages of anabolism.

- Stage 1 involves production of precursors such as amino acids, monosaccharides, isoprenoids and nucleotides.

- Stage 2 involves activation of these precursors into reactive forms using energy from ATP.

- Stage 3 involves the assembly of these precursors into complex molecules such as proteins, polysaccharides, lipids and nucleic acids.

Sources of Energy for Anabolic Processes

Different species of organisms depend on different sources of energy. Autotrophs such as plants can construct the complex organic molecules in cells such as polysaccharides and proteins from simple molecules like carbon dioxide and water using sunlight as energy.

Heterotrophs, on the other hand, require a source of more complex substances, such as monosaccharides and amino acids, to produce these complex molecules. Photoautotrophs and photoheterotrophs obtain energy from light while chemoautotrophs and chemoheterotrophs obtain energy from inorganic oxidation reactions.

Anabolism of Carbohydrates

In these steps simple organic acids can be converted into monosaccharides such as glucose and then used to assemble polysaccharides such as starch. Glucose is made from pyruvate, lactate, glycerol, glycerate 3-phosphate and amino acids and the process is called gluconeogenesis. Gluconeogenesis converts pyruvate to glucose-6-phosphate through a series of intermediates, many of which are shared with glycolysis.

Usually fatty acids stored as adipose tissues cannot be converted to glucose through gluconeogenesis as these organisms cannot convert acetyl-CoA into pyruvate. This is the reason why when there is long term starvation, humans and other animals need to produce ketone bodies from fatty acids to replace glucose in tissues such as the brain that cannot metabolize fatty acids.

Plants and bacteria can convert fatty acids into glucose and they utilize the glyoxylate cycle, which bypasses the decarboxylation step in the citric acid cycle and allows the transformation of acetyl-CoA to oxaloacetate. From this glucose is formed.

Glycans and polysaccharides are complexes of simple sugars. These additions are made possible by glycosyltransferase from a reactive sugar-phosphate donor, such as uridine diphosphate glucose (UDP-glucose), to an acceptor hydroxyl group on the growing polysaccharide. The hydroxyl groups on the ring of the substrate can be acceptors and thus polysaccharides produced can have straight or branched structures. These polysaccharides so formed may be transferred to lipids and proteins by enzymes called oligosaccharyltransferases.

Anabolism of Proteins

Proteins are formed of amino acids. Most organisms can synthesize some of the 20 common amino acids. Most bacteria and plants can synthesize all twenty, but mammals can synthesize only the ten nonessential amino acids.

The amino acids are joined together in a chain by peptide bonds to form polypeptide chains. Each different protein has a unique sequence of amino acid residues: this is its primary structure. The polypeptide chain undergoes modifications, folding and structural changes to form the final protein.

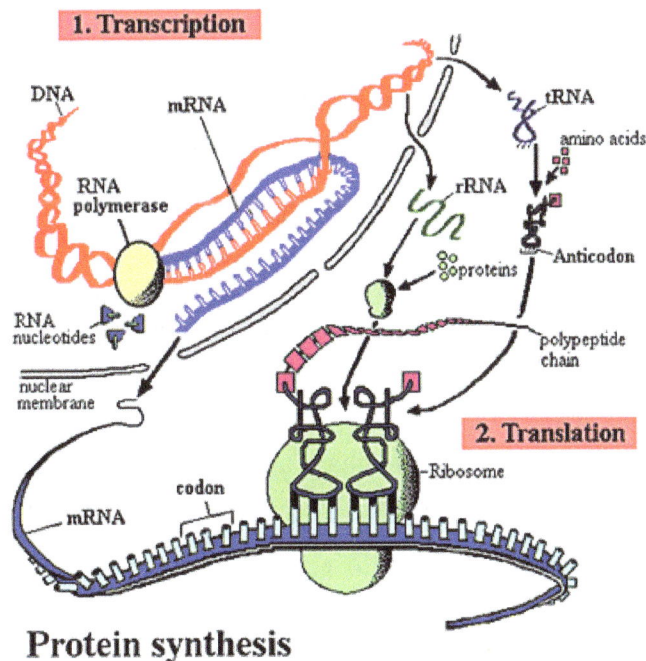

Protein synthesis

Nucleotides are made from amino acids, carbon dioxide and formic acid in pathways that require large amounts of metabolic energy.

Purines are synthesized as nucleosides (bases attached to ribose). Adenine and guanine for example are made from the precursor nucleoside inosine monophosphate, which is synthesized using atoms from the amino acids glycine, glutamine, and aspartic acid, as well as formate transferred from the coenzyme tetrahydrofolate.

Pyrimidines, like thymine and cytosine, are synthesized from the base orotate, which is formed from glutamine and aspartate.

Anabolism of Fatty Acids

Fatty acids are synthesized using fatty acid synthases that polymerize and then reduce acetyl-CoA units. These fatty acids contain acyl chains that are extended by a cycle of reactions that add the actyl group, reduce it to an alcohol, dehydrate it to an alkene group and then reduce it again to an alkane group.

In animals and fungi, all these fatty acid synthase reactions are carried out by a single multifunctional type I protein. In plants, plasmids and bacteria separate type II enzymes perform each step in the pathway.

Other lipids like terpenes and isoprenoids include the carotenoids and form the largest class of plant natural products. These compounds are made by the assembly and modification of isoprene units donated from the reactive precursors isopentenyl pyrophosphate and dimethylallyl pyrophosphate. In animals and archaea, the mevalonate pathway produces these compounds from acetyl-CoA.

Bacterial Proteins

A bacterial protein is a protein which is either part of the structure of a bacterium, or produced by a bacterium as part of its life cycle. Proteins are an important part of all living organisms, and bacteria are no exception. Thanks to the fact that many bacteria are easy to culture in the laboratory, a great deal of research on bacterial proteins has been performed with the goal of learning more about specific proteins and their functions. Understanding bacterial proteins is important both because bacteria play a very active role in human health, and because the information can be extrapolated to gather more data about the proteins associated with larger organisms, including humans.

Proteins are lengthy chains of amino acids which are folded back upon themselves. The nature of a protein is determined both by the amino acid chain and by the way in which the protein is folded. Proteins are encoded in the genes, with certain proteins being expressed while an organism develops, with others are produced by an organism with the goal of accomplishing specific tasks. The genetic code of an organism holds the blueprints for numerous proteins.

In addition to being a unique structure, a bacterial protein also has the ability to bind with other proteins. Protein binding involves the formation of very strong links between two different proteins. Once proteins bind, they can trigger a reaction which may vary from an immune system response to an infection to the onset of a disease. Over time, many bacteria have evolved to produce proteins which target particular locations on human and animal cells.

Bacterial proteins are of interest to humans for a number of reasons. Understanding which proteins are involved in the structure of particular bacteria can help researchers develop medications which identify and target a particular bacterial protein, allowing the researchers to create antibacterial drugs which target specific organisms. Understanding individual proteins can also allow researchers to monitor mutations and to keep track of the ways in which these mutations occurred, and how they can be addressed.

Some bacteria produce proteins which have a deleterious effect on the human body. A bacterial protein can be toxic, causing illness or death in an organism which has been infected by the bacteria, and bacterial proteins can also bind with specific proteins in the body to cause a variety of symptoms. Researchers can spend years identifying all of the proteins associated with a single type of bacterium, and this process can be complicated by rapid mutations, as seen in the case of the wily Staphylococcus bacterium.

Carbohydrates

All bacteria must utilize the energy sources in their environment in order to produce ATP. ATP is required for all of the biosynthetic processes that bacteria use for their maintenance and reproduction. Bacteria produce enzymes that allow them to oxidize environmental energy sources; however, the energy sources that different bacteria use depends on the specific enzymes that each bacteria produce.

Heterotrophic bacteria often use carbohydrates as energy sources. Many bacteria use glucose, a monosaccharide or simple sugar, because many bacteria possess the enzymes required for the degradation and oxidation of this sugar. Fewer bacteria are able to use complex carbohydrates like disaccharides (lactose or sucrose) or polysaccharides (starch). Disaccharides and polysaccharides are simple sugars that are linked by glycosidic bonds; bacteria must produce enzymes to cleave these bonds such that the simple sugars that result can be transported into the cell. If the bacteria cannot produce these enzymes then the complex carbohydrate is not used. For example, lactose is a disaccharide consisting of monomeric glucose and monomeric galactose linked by a glycosidic bond. Bacteria that use lactose must first produce the enzyme lactase (beta-galactosidase) to break the glycosidic bond between these monomers. Starch is a large polysaccharide consisting of long chains of monomeric glucose linked by glycosidic bonds. Bacteria that use starch produce an exoenzyme, alpha amylase, that break these bonds such that free monomeric glucose is produced.

Each bacterium has its own collection of enzymes that enable it to use diverse carbohydrates; this is often exploited in the identification of bacterial species. One can determine if a given bacterial species can utilize a given carbohydrate by checking for the presence of byproducts that are produced by the oxidation of these carbohydrates. To this end, pH indicators may be added to the media to detect metabolic acids that have been produced by bacteria after the oxidation and fermentation of sugars. (Phenol Red with Durham tubes, Citrate agar slants) Alternatively, reagents may be added to media after the bacteria have grown, these reagents react with specific byproducts or intermediates in a metabolic pathway.

Glucose or Lactose Broths with Durham Tubes

These media contain peptone as an alternative energy source and either glucose or lactose as a carbohydrate source. The glucose or lactose is added to detect gas production and metabolic acid production resulting from bacterial utilization of the carbohydrate present in each medium. Both media contain phenol red that is a pH indicator for the presence of organic acids. If the bacteria utilize either glucose or lactose and produce organic acids then the pH indicator will turn yellow in color. Durham tubes are submerged in the medium to detect gas production. Durham tubes are inverted (top of tube facing the bottom of the test tube) such that the tubes are closed at the top and open at the bottom. When gas is produced in these media a small bubble representing trapped gas can be found in the tube.

There are four possible reactions in glucose or lactose broths with Durham tubes:

1. No reaction (-) with no growth: The broth remains red in color—the organism can not utilize the carbohydrate under the relatively anaerobic conditions of the medium and did not grow.

2. No reaction (-) with growth: The broth remains red in color but the medium becomes tur-bid or cloudy—the organism can grow in the media but is not using the carbohydrates; these organisms are using the supplied peptone as the energy source.

3. Acid production (A): The broth turns yellow in color--the organism is able to use the carbo-hydrate and produces organic acids that lower the pH of the medium.

4. Acid and Gas production (A/G): The broth turns yellow in color and a small bubble is present in the inverted Durham tube—the organism is able to use the carbohydrate and produced organic acids and gas in the process.

Methyl Red—vogues Proskauer (Mrvp) Medium

MRVP media is primarily used for the identification of enteric bacteria like Escherichia coli, Shi-gella and Salmonella spp. and Enterobacter aerogenes. MRVP media is used to determine the pathway a given organism uses to ferment glucose; to this end reagents such as methyl red (a pH indicator) and alpha-napthol and potassium hydroxide are added to detect end products or inter-mediates produced by a given fermentation pathway.

After the bacteria have grown in the MRVP media, the media is split to perform two separate tests—the Methyl red test and the Vogues Proskauer test. The MR test employs the addition of the pH indicator methyl red, to determine whether an organism has produced high levels of acid during glucose fermentation. Bacteria that use the mixed acid fermentation pathway produce high levels of organic acid—such high levels of acid causes the methyl red to turn red in color. (NOTE: This is to be distinguished from phenol red!!) If the amount of acid production is relatively lower, then the medium remains yellow or turns slightly orange after the addition of methyl red. If the media remains red this is a positive (+) reaction for the MR test if the medium remains yellow or turns slightly orange, this is a negative reaction for the MR test.

The VP test is used to detect acetoin. Bacteria that use the butandiol fermentation pathway (as opposed to the mixed acid fermentation pathway) also produce acetoin that is an intermediate of this pathway. One adds alpha-napthol and potassium hydroxide to the divided medium and periodically shakes the medium to introduce oxygen to the reaction. After 20 minutes if acetoin is present, the medium will turn pink—this is a positive reaction.

There are four possible reactions one can observe with the MR/VP tests

+/-   -/+   +/+   -/-

Simmons Citrate Agar

Simmons citrate agar is used to determine if bacteria are able to utilize citrate as a sole carbon source. Citrate (citric acid) is an intermediate of the Kreb's cycle (citric acid cycle) used in respi-ratory metabolism. A pH indicator, brom thymol blue is incorporated into the medium. Brom thymol blue is green at pH 7 and turns blue at a pH above 7.6. Therefore, if the citrate (citric acid) has been used as a sole carbon source, then the pH of the medium will increase imparting a blue color to the pH indicator.

There are two possible reactions that can be observed with the Simmons Citrate Agar.

1. Citrate (+) –the medium turns blue: The bacteria have used citric acid as a sole carbon source, thus raising the pH of the medium.

2. Citrate (-) –the medium remains green: The bacteria are not able to use citric acid as a sole carbon source; the pH of the medium remains the same.

Nucleotides

Nucleotides are essential for many cellular functions, including the storage of genetic information, gene expression, energy metabolism, cell signaling, and biosynthesis. In a cell, nucleotides exist primarily as 5'-triphosphates. ATP is the most prevalent nucleotide, reaching mM concentrations in many cell types, while other nucleotides may be present at much lower concentrations (cAMP).

Biological Significance of Nucleotides

1. Building blocks of nucleic acids (DNA and RNA).

2. Involved in energy storage, muscle contraction, active transport, and maintenance of ion gradients.

3. Activated intermediates in biosynthesis (e.g. UDP-glucose, S-adenosylmethionine (SAM).

4. Components of coenzymes (NAD^+, $NADP^+$, FAD, FMN, and CoA)

5. Metabolic regulators:

    a. Second messengers (cAMP, cGMP)

    b. Phosphate (PO_3^{2-}) donors for phosphorylation of kinases and phosphatases in signal transduction (ATP)

    c. Regulation of some enzymes via adenylation and uridylylation Each nucleotide contains a purine or pyrimidine base, a ribose or deoxyribose sugar, and a phosphate:

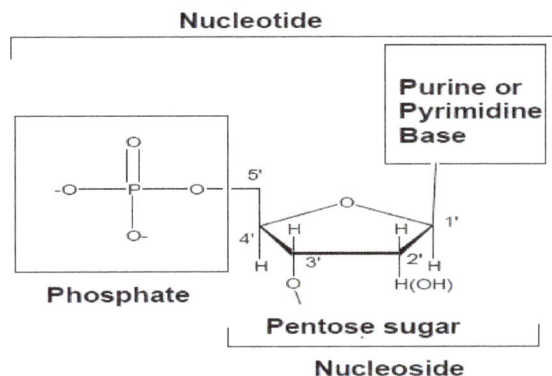

Figure: Strcuture of Nucleotide.

Sources of Nucleotides in a Cell

Degradation of nucleic acids (salvage pathways). Free purine bases can be recycled by coupling with the ribose phosphate moiety, 5-phospho-ribosyl-1-pyrophosphate (PRPP), to form nucleotide monophosphates:

adenine phosphoribosyl
transferase

Adenine + PRPP ⟶ Adenylate + PPi

Hypoxanthine-guanine phosphoribosyl
transferase (HGPRT)

Guanine + PRPP   Guanylate + PPi

Hypoxanthine + PRPP ⟶ Inosinate+ PPi

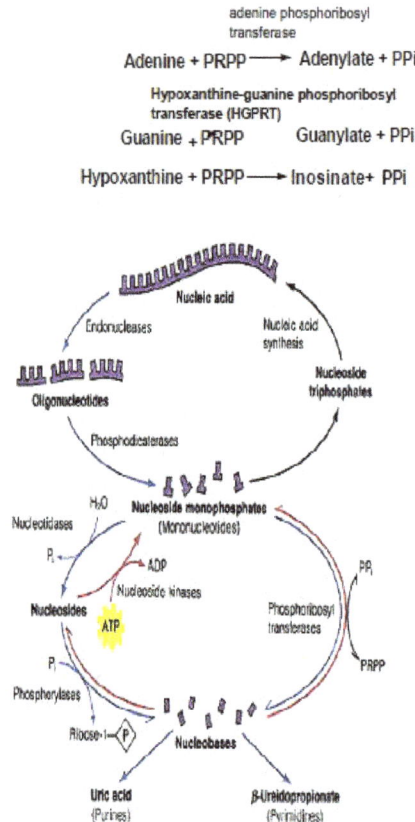

Figure: Soruces of Nucleotides.

Minerals

The ability of living organisms to form minerals is the fundamental tenet of biomineralogy. Among plants and animals, this process involves the production of cystolith inclusions in leaves and hard mineralized body parts like bones, teeth, and shells. This process, biological mineral precipitation, is not exclusive to higher eukaryotic organisms. Prokaryotic microorganisms, or bacteria, are remarkably potent agents of biomineralization, too. These small wonders manage to form an enormous variety of minerals, carbonates, phosphates, oxides, sulfides, and silicates as well as silver and gold.

Microbial biomineralogy has extremely broad and deep biogeochemical roots. Bacteria are the most abundant and metabolically diverse forms of life on Earth. They grow under a wide range of geochemical conditions in an unparalleled variety of habitats. Basically, microbial life exists wherever there is liquid water at temperatures from -7°C to about 120°C. Even the most extreme environments, from Antarctica to the ocean bottom and deep underground, play host to thriving microbial populations. Fossil and isotopic evidence also reveal that microbial life is as old as the rock records, stretching back at least 3.8 billion years.

How Minerals Develop

Microorganisms produce minerals in two distinct ways, passive growth and as a result of metabolic activity. The first process involves the nucleation and growth of crystals from an oversaturated solution on the outside surface of individual cells. This happens because the cell walls and external sheaths of bacterial cells have an abundance of chemically reactive sites that bind dissolved mineral-forming elements. When this adsorption occurs, the activation energy barrier that normally inhibits spontaneous nucleation and crystal growth is greatly reduced. Epicellular mineral precipitation follows, often leading to the complete encrustation of cells. Bacterial precipitation of amorphous silica in hot springs provide good examples of this type of microbial biomineralization. Some forms of authigenic iron oxides, phosphates, carbonates, and clays develop in the way.

Microbial mineral precipitation also results from metabolic activities of microorganisms. The process can occur inside or outside the cells, or even some distance away. Often, bacterial activity simply triggers a change in solution chemistry that leads to oversaturation and mineral precipitation. For example, the growth of photosynthetic cyanobacteria in natural alkaline waters tends to promote an increase in pH. This supports the precipitation of carbonate minerals like calcite and strontianite. Similarly, sulfide production of mackinawite, pyrite, and other sulfide minerals occurs, particularly in marine sediments where these bacteria flourish.

Other mineral phases precipitate directly from bacterial enzyme action. Enzymes are proteins that catalyze chemical reactions and drive cellular metabolism. For example, diverse forms of iron and manganese oxides are deposited by bacteria that actively oxidize soluble, reduced forms of the metals to generate energy for growth. The formation of tiny magnetite particles inside magneto-tactic bacteria, and the reductive precipitation of uraninite by some metal-reducing bacteria, are further examples of enzymatically formed minerals. Bacterial formation of metallic gold and silver might be related, too, as it must stem from some kind of reductive precipitation. But with these metals, it isn't clear if enzymes are involved.

Small Grain Sizes

Regardless of how they are formed, mineral precipitates produced by microorganisms usually have an extremely small grain size and often exist in a poorly ordered, near amorphous state. This may be related to high rates of nucleation and precipitation. In some cases, mineral precipitates are fine enough to preserve microbial cell structure. Silica is a good example, forming small crystallites that cause complete silicification of structurally intact cells. Paleont-ologists study the mineralization of modern microbial cells by silica to gain insight into how ancient microorganisms were preserved as fossils billions of years ago in silicified carbonates and cherts.

The small grain size of microbial mineral precipitates also confers high surface reactivity; these minerals act as secondary adsorbents of dissolved inorganic cations and anions, and possibly even organic compounds. Whether this process benefits or harms microorganisms is not yet known. But we do know that the chemical composition of natural bodies of water is strongly influenced by the adsorption of dissolved substances to suspended particulate materials. These particulates are often made of living and dead bacteria encrusted with fine-grain minerals, including manganese and iron oxides. The implication is that microbial mineral precipitation helps regulate the chemistry of aquatic systems.

Geochemical Cycling

On a global environmental scale, microbial biomineralogy plays a major role in the geochemical cycling of mineral-forming elements. The transfer of dissolved iron and sulfur into marine sediments, for example, is driven mainly by microbial pyrite precipitation. Microbiological precipitation of phosphate minerals also contributes to the incorporation of phosphorus into sediments, particularly in oceanic upwelling zones along the west coasts of South America and Africa. Even the chemical weathering of continental rocks and the atmosphere are influenced by microbial biomineralogy. In this instance, the precipitation of carbonate minerals by microorganisms is especially relevant because these minerals serve as a sink for atmospheric carbon dioxide and as end members in the weathering of silicate minerals, such as feldspars or olivines of igneous rocks.

Oxidative Phosphorylation

Oxidative phosphorylation, i.e., the formation of energy-rich adenosine triphosphate from the energy of substances oxidized in the organism, is generally recognized as one of the most important processes in the cells of animals and most microorganisms, since it ensures the accomplishment of almost all the other vital processes. The term "oxidative phosphorylation," as distinct from photosynthetic and glycolytic phosphorylation, means that the transfer of electrons (hydrogen) from the oxidizable substance to oxygen or another acceptor is effected by means of specific dehydrogenases and other enzymes forming the respiratory chain.

The Electron Transport Chain

The electron transport chain is a collection of membrane-embedded proteins and organic molecules, most of them organized into four large complexes labeled I to IV. In eukaryotes, many copies of these molecules are found in the inner mitochondrial membrane. In prokaryotes, the electron transport chain components are found in the plasma membrane.

As the electrons travel through the chain, they go from a higher to a lower energy level, moving from less electron-hungry to more electron-hungry molecules. Energy is released in these "downhill" electron transfers, and several of the protein complexes use the released energy to pump protons from the mitochondrial matrix to the intermembrane space, forming a proton gradient.

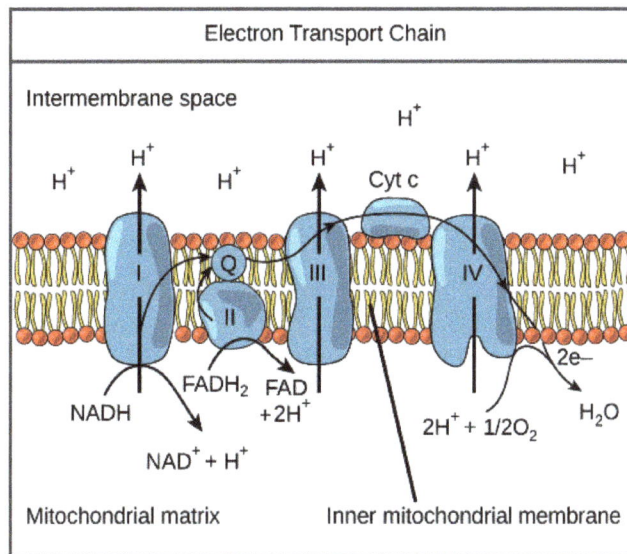

Figure: Oxidative phosphorylation

All of the electrons that enter the transport chain come from NADH and $FADH_2$ molecules produced during earlier stages of cellular respiration: glycolysis, pyruvate oxidation, and the citric acid cycle.

- NADH is very good at donating electrons in redox reactions (that is, its electrons are at a high energy level), so it can transfer its electrons directly to complex I, turning back into NAD^+. As electrons move through complex I in a series of redox reactions, energy is released, and the complex uses this energy to pump protons from the matrix into the intermembrane space.

- $FADH_2$ is not as good at donating electrons as NADH (that is, its electrons are at a lower energy level), so it cannot transfer its electrons to complex I. Instead, it feeds them into the transport chain through complex II, which does not pump protons across the membrane.

Because of this "bypass," each $FADH_2$ molecule causes fewer protons to be pumped (and contributes less to the proton gradient) than an NADH.

Beyond the first two complexes, electrons from NADH and $FADH_2$ travel exactly the same route. Both complex I and complex II pass their electrons to a small, mobile electron carrier called ubiquinone (Q), which is reduced to form QH_2 and travels through the membrane, delivering the electrons to complex III. As electrons move through complex III, more H^+ ions are pumped across the membrane, and the electrons are ultimately delivered to another mobile carrier called cytochrome C (cyt C). Cyt C carries the electrons to complex IV, where a final batch of H^+ ions is pumped across the membrane. Complex IV passes the electrons to O_2, which splits into two oxygen atoms and accepts protons from the matrix to form water. Four electrons are required to reduce each molecule of O_2, and two water molecules are formed in the process.

Overall, what does the electron transport chain do for the cell? It has two important functions:

- Regenerates electron carriers: NADH and $FADH_2$ pass their electrons to the electron transport chain, turning back into NAD^+ and FAD. This is important because the oxidized forms of these electron carriers are used in glycolysis and the citric acid cycle and must be available to keep these processes running.

- Makes a proton gradient: The transport chain builds a proton gradient across the inner mitochondrial membrane, with a higher concentration of H^+ in the intermembrane space and a lower concentration in the matrix. This gradient represents a stored form of energy, and, as we'll see, it can be used to make ATP.

Chemiosmosis

Complexes I, III, and IV of the electron transport chain are proton pumps. As electrons move energetically downhill, the complexes capture the released energy and use it to pump H^+ ions from the matrix to the intermembrane space. This pumping forms an electrochemical gradient across the inner mitochondrial membrane. The gradient is sometimes called the proton-motive force, and you can think of it as a form of stored energy, kind of like a battery.

Like many other ions, protons can't pass directly through the phospholipid bilayer of the membrane because its core is too hydrophobic. Instead, H^+ ions can move down their concentration gradient only with the help of channel proteins that form hydrophilic tunnels across the membrane.

In the inner mitochondrial membrane, H^+ ions have just one channel available: a membrane-spanning protein known as ATP synthase. Conceptually, ATP synthase is a lot like a turbine in a hydroelectric power plant. Instead of being turned by water, it's turned by the flow of H^+ ions moving down their electrochemical gradient. As ATP synthase turns, it catalyzes the addition of a phosphate to ADP, capturing energy from the proton gradient as ATP.

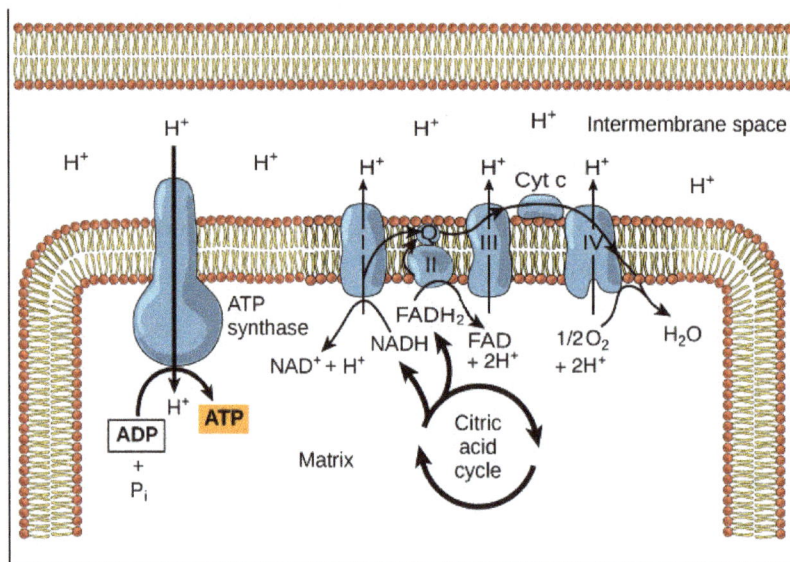

This process, in which energy from a proton gradient is used to make ATP, is called chemiosmosis. More broadly, chemiosmosis can refer to any process in which energy stored in a proton gradient is used to do work. Although chemiosmosis accounts for over 80% of ATP made during glucose breakdown in cellular respiration, it's not unique to cellular respiration. For instance, chemiosmosis is also involved in the light reactions of photosynthesis.

What would happen to the energy stored in the proton gradient if it weren't used to synthesize ATP or do other cellular work? It would be released as heat, and interestingly enough, some types of cells deliberately use the proton gradient for heat generation rather than ATP synthesis. This might seem wasteful, but it's an important strategy for animals that need to keep warm. For instance, hibernating mammals (such as bears) have specialized cells known as brown fat cells. In the brown fat cells, uncoupling proteins are produced and inserted into the inner mitochondrial membrane. These proteins are simply channels that allow protons to pass from the intermembrane space to the matrix without traveling through ATP synthase. By providing an alternate route for protons to flow back into the matrix, the uncoupling proteins allow the energy of the gradient to be dissipated as heat.

ATP Yield

How many ATP do we get per glucose in cellular respiration? If you look in different books, or ask different professors, you'll probably get slightly different answers. However, most current sources estimate that the maximum ATP yield for a molecule of glucose is around 30-32 ATP[2,3,4] This range is lower than previous estimates because it accounts for the necessary transport of ADP into, and ATP out of, the mitochondrion.

Where does the figure of 30-32 ATP come from? Two net ATP are made in glycolysis, and another two ATP (or energetically equivalent GTP) are made in the citric acid cycle. Beyond those four, the remaining ATP all come from oxidative phosphorylation. Based on a lot of experimental work, it appears that four H^+ ions must flow back into the matrix through ATP synthase to power the synthesis of one ATP molecule. When electrons from NADH move through the transport chain, about 10 H^+ ions are pumped from the matrix to the intermembrane space, so each NADH yields about 2.5 ATP. Electrons from $FADH_2$ which enter the chain at a later stage, drive pumping of only 6 H^+ leading to production of about 1.5 ATP.

With this information, we can do a little inventory for the breakdown of one molecule of glucose:

Stage	Direct products (net)	Ultimate ATP yield (net)
Glycolysis	2 ATP	2 ATP
	2 NADH	3-5 ATP
Pyruvate oxidation	2 NADH	5 ATP
Citric acid cycle	2 ATP/GTP	2 ATP
	6 NADH	15 ATP
	2 $FADH_2$	3 ATP
Total		30-32 ATP

One number in this table is still not precise: the ATP yield from NADH made in glycolysis. This is because glycolysis happens in the cytosol, and NADH can't cross the inner mitochondrial membrane to deliver its electrons to complex I. Instead, it must hand its electrons off to a molecular "shuttle system" that delivers them, through a series of steps, to the electron transport chain.

- Some cells of your body have a shuttle system that delivers electrons to the transport chain via $FADH_2$. In this case, only 3 ATP are produced for the two NADH of glycolysis.

- Other cells of your body have a shuttle system that delivers the electrons via NADH, resulting in the production of 5 ATP. In bacteria, both glycolysis and the citric acid cycle happen in the cytosol, so no shuttle is needed and 5 ATP are produced.

30-32 ATP from the breakdown of one glucose molecule is a high-end estimate, and the real yield may be lower. For instance, some intermediates from cellular respiration may be siphoned off by the cell and used in other biosynthetic pathways, reducing the number of ATP produced. Cellular respiration is a nexus for many different metabolic pathways in the cell, forming a network that's larger than the glucose breakdown pathways alone.

References

- Bacterial-physiology-and-metabolism: encyclopedia2.thefreedictionary.com, Retrieved 14 May 2018

- What-is-Anabolism: news-medical.net, Retrieved 23 March 2018

- Microbes-minerals: uwaterloo.ca, Retrieved 25 June 2018

- Oxidative-phosphorylation, cellular-respiration-and-fermentation: khanacademy.org, Retrieved 15 April 2018

- What-is-a-bacterial-protein: wisegeek.com, Retrieved 11 July 2018

Chapter 4

Bacterial Genetics and Growth

Bacteria have a single chromosome, plasmids, extra-chromosomal DNAs that consist of genes that are responsible for metabolism, antibiotic resistance and virulence. Bacterial genomes encode between a hundred to a thousand genes. Bacteria grow to maturity and reproduce through binary fission. This chapter discusses the genetics and growth in bacteria through an analysis of bacterial conjugation, measurement of bacterial growth, growth curve, etc.

Bacterial Genetics

Bacterial genetics is the study of the mechanisms of heritable information in bacteria, their chromosomes, plasmids, transposons and phages. Techniques that have enabled this discipline are culture in defined media, replica plating, mutagenesis, transformation, conjugation and transduction.

Beneficial mutations that develop in one bacterial cell can also be passed to related bacteria of different lineages through the process of horizontal transmission. There are three main forms of horizontal transmission used to spread genes between members of the same or different species: conjugation (bacteria-to-bacteria transfer), transduction (viral-mediated transfer), and transformation (free DNA transfer). These forms of genetic transfer can move plasmid , bacteriophage, or genomic DNA sequences. A plasmid is a small circle of DNA separate from the chromosome; a bacteriophage is a virus that reproduces in bacteria by injecting its DNA; the genome is the total DNA of the bacterial organism.

After transfer, the DNA molecules can exist in two forms, either as DNA molecules separate from the bacterial chromosome (an episome), or can become part of the bacterial chromosome. The study of basic mechanisms used by bacteria to exchange genes allowed scientists to develop many of the essential tools of modern molecular biology.

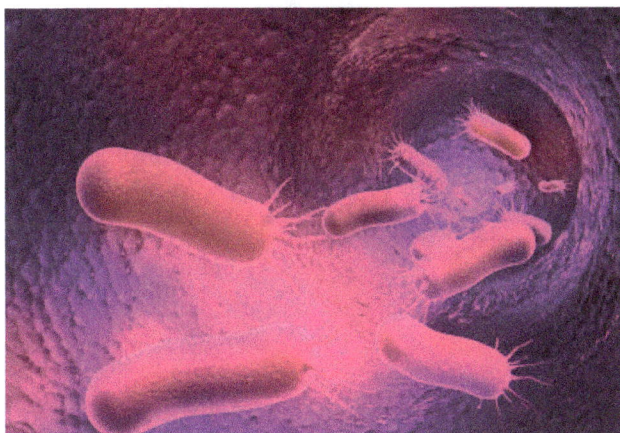

Conjugation

Bacterial conjugation refers to the transfer of DNA between bacterial cells that requires cell-to-cell contact. Joshua Lederberg and Edward Tatum first described conjugation in 1946 when they discovered the F factor (an episome) that can move between Escherichia coli cells. The F factor is one of the most well studied conjugative plasmids (plasmids are circular episomes) and is the most well studied conjugative system. There are many different conjugal plasmids carried by members of most bacterial species. Conjugal plasmids that carry antibiotic resistance genes are called R factors. The F factor and R factors usually exist as episomes and each carries functions that allow it to replicate its DNA and thus be inherited by the daughter cells after binary fission. However, conjugative plasmids also express transfer functions that allow the movement of DNA from a donor to a recipient cell; this is the process of conjugation.

The steps of bacterial conjugation are: mating pair formation, conjugal DNA synthesis, DNA transfer, and maturation. The main structure of the F factor that allows mating pair formation is the F pilus or sex pilus (a long thin fiber that extends from the bacterial cell surface). There are one to three pili expressed on an E. coli cell that carries the F factor, and one pilus will specifically interact with several molecules on the recipient cell surface (attachment). About twenty genes on the F factor are required to produce a functional pilus, but the structure is mainly made up of one protein , pilin. To bring the donor and recipient cell into close proximity, the F pilus retracts into the donor cell by removing pilin protein monomers from the base of the pilus to draw the bacterial cells together.

Once a stable mating pair is formed, a specialized form of DNA replication starts. Conjugal DNA synthesis produces a single-stranded copy of the F factor DNA (as opposed to a double-stranded DNA that is formed by normal replication). This DNA strand is transferred into the recipient cell. Once in the recipient cell, the single-stranded copy of the F plasmid DNA is copied to make a double-stranded DNA molecule, which then forms a mature circular plasmid. At the end of conjugation the mating pair is broken and both the donor and the recipient cells carry an identical episomal copy of the F factor. All of the approximately one hundred genes carried on the F factor can now be expressed by the recipient cell and will be inherited by its offspring.

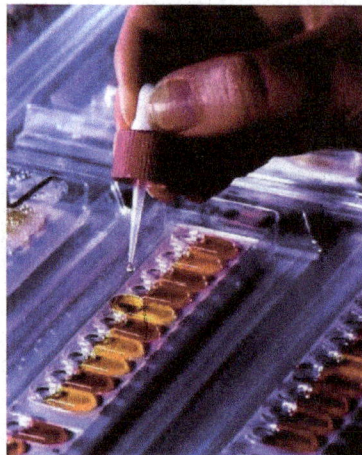

A laboratory technician performing an Analytical Profile Index (API) test on bacteria.

In addition to transferring itself, the F factor can also transfer chromosomal genes between a donor and recipient cell. The F factor can be found inserted (integrated) into the bacterial chromosome at many locations in a small fraction of bacterial cells. An integrated F factor is replicated along with the rest of the chromosome and inherited by offspring along with the rest of the chromosome. When a mating pair is formed between the donor cell carrying an integrated F factor and a recipient cell, DNA transfer occurs as it does for the episomal F factor, but now the chromosomal sequences adjacent to the integrated F factor are transferred into the recipient. Since these DNA sequences encode bacterial genes, they can recombine with the same genes in the recipient. If the donor gene has minor changes in DNA sequence from the recipient gene, the different sequence can be incorporated into the recipient gene and inherited by the recipient cell's offspring. Donor cells that have an integrated copy of the F factor are called Hfr strains (High frequency of recombination).

Transduction

The second way that DNA is transferred between bacterial cells is through a phage particle in the process of transduction. Joshua Lederberg and Norton Zinder first discovered transduction in 1956. When phage inject their DNA into a recipient cell, a process occurs that produces new bacteriophage particles and kills the host cell (lytic growth). Some phage do not always kill the host cell (temperate phage), but instead can be inherited by daughter host cells. Therefore acquisition of a so-called temperate "prophage" by a recipient cell is a form of transduction. Many phage also have the ability to transfer chromosomal or plasmid genes between bacterial cells. During generalized transduction any gene can be transferred from a donor cell to a recipient cell. Generalized transducing phage are produced when a phage packages bacterial genes into its capsid (protein envelope) instead of its own DNA. When a phage particle carrying bacterial chromosomal genes attaches to a recipient cell, the DNA is injected into the cytoplasm where it can recombine with a homologous DNA sequences.

Some bacteriophage can pick up a subset of chromosomal genes and transfer them to other bacteria. This process is called specialized transduction since only a limited set of chromosomal genes can be transferred between bacterial cells.

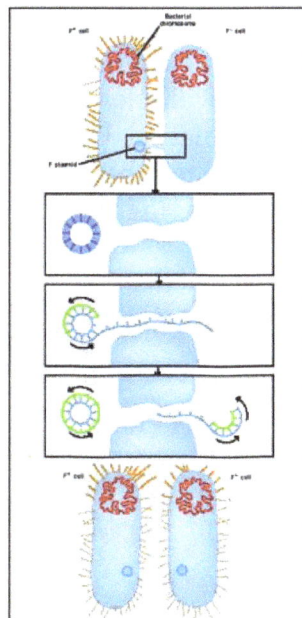

Bacterial conjugation. The bacterium on the left passes a copy of the F plasmid to the bacterium on the right, converting it from an F - cell to an F + cell.

The third main way that bacteria exchange DNA is called DNA transformation. Some bacteria have evolved systems that transport free DNA from the outside of the bacterial cell into the cytoplasm. These bacterial are called "naturally competent" for DNA transformation. Natural DNA transformation of *Streptococcus pneumonaiae* provided the first proof that DNA encoded the genetic material in experiments by Oswald Avery and colleagues. Some other naturally competent bacteria include *Bacillus subtilis*, *Haemophilus influenzae*, and *Neisseria gonorrhoeae*. Other bacterial species such as *E. coli* are not naturally competent for DNA transformation. Scientists have devised many ways to physically or chemically force noncompetant bacteria to take up DNA. These methods of artificial DNA transformation form the basis of plasmid cloning in molecular biology.

Most naturally competent bacteria regulate transformation competence so that they only take up DNA into their cells when there is a high density of cells in the environment. The ability to sense how many other cells are in an area is called quorum sensing. Bacteria that are naturally competent for DNA transformation express ten to twenty proteins that form a structure that spans the bacterial cell envelope. In some bacteria this structure also is required to form a particular type of pilus different than the F factor pilus. Other bacteria express similar structures that are involved in secreting proteins into the exterior medium (Type II secretion). Therefore, it appears that DNA transformation and protein secretion have evolved together.

During natural DNA transformation, doubled-stranded DNA is bound to the recipient cell surface by a protein receptor. One strand of the DNA is transported through the cell envelope, where it can recombine with similar sequences present in the recipient cell. If the DNA taken up is not homologous to genes already present in the cell, the DNA is usually broken down and the nucleotides released are used to synthesize new DNA during normal replication. This observation has led to the speculation that DNA transformation competence may have originally evolved to allow the acquisition of nucleic acids for food.

The source of DNA for transformation is thought to be DNA released from other cells in the same population. Most naturally competent bacteria spontaneously break apart by expressing enzymes that break the cell wall. Autolysis will release the genomic DNA into the environment where it will be available for DNA transformation. Of course, this results in the death of some cells in the population, but usually not large numbers of cells. It appears that losing a few cells from the population is counterbalanced by having the possibility of gaining new traits by DNA transformation.

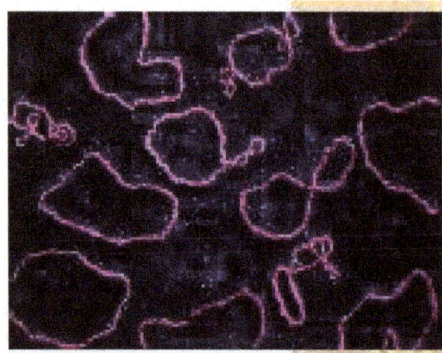

A scanning electron micrograph of bacterial DNA plasmids.

Bacterial Growth

Bacterial growth is finely tuned to cellular state and environmental input. Growth not only determines the fate of a single cell, but also of the community, and both depends on and is regulated by other cells, including kin cells, other microorganisms and host cells. What governs bacterial growth is a classic question of microbiology with a rich history of research, and modern tools and approaches have invigorated the field.

Bacteria replicate by binary fission, a process by which one bacterium splits into two. Therefore, bacteria increase their numbers by geometric progression whereby their population doubles every generation time. Generation time is the time it takes for a population of bacteria to double in number. For many common bacteria, the generation time is quite short, 20-60 minutes under optimum conditions. For most common pathogens in the body, the generation time is probably closer to 5-10 hours. Because bacteria grow by geometric progression and most have a short generation time, they can astronomically increase their number in a short period of time.

The relationship between the number of bacteria in a population at a given time (Nt), the original number of bacterial cells in the population (No), and the number of divisions those bacteria have undergone during that time (n) can be expressed by the following equation:

$N_t = N_o \times 2^n$

For example, Escherichia coli, under optimum conditions, has a generation time of 20 minutes. If one started with only 10 E. coli (No = 10) and allowed them to grow for 12 hours (n = 36; with a generation time of 20 minutes they would divide 3 times in one hour and 36 times in 12 hours), then plugging the numbers in the formula, the number of bacteria after 12 hours (Nt) would be,

$10 \times 2^{36} = N_t = 687,194,767,360 \, E.coli$

In general it is thought that during DNA replication, each strand of the replicating bacterial DNA attaches to proteins at what will become the cell division plane. For example, Par proteins function to separate bacterial chromosomes to opposite poles of the cell during cell division. They bind to the origin of replication of the DNA and physically pull or push the chromosomes apart, similar to the mitotic apparatus of eukaryotic cells.

Physical Conditions Required for Growth

The physical environment in which the organisms will grow best. They exhibit diverse responsive to physical conditions such as Temperature, Gaseous conditions, and pH.

Temperature

The best temperature for bacterial growth is different. If some bacteria grows in artic oceans of hot springs, it is not surprising matter. The optimum temperature of bacteria is not constant. In human GUT some bacteria grows very well at human body temperature 37°C. But plant bacteria are not able to survive at those temperatures.

All processes of growth are dependent on chemical reactions and since the rates of those reactions are influenced by temperature. The temperature that allows for most rapid growth during a short period of time (12 to 24 hours) is known as the "Optimum temperature".

On the basis of their temperature relationships, Bacteria are divided into three main groups:

Psychrophiles

- These are able to grow at 0°C (or) lower, though they grow best at a higher temperature.

- Many microbiologists restrict the term Psychrophile to organisms that can grow at 0°C but have an optimum temperature of 15°C (or) lower and maximum temperature of about 20°C.

- Facultative psychrophile organisms are able to grow at 0°C; and temperature in the range of about 20 to 30°C.

Mesophiles

- These grow best within a temperature range of approximately 25 to 40°C.E.g.: Pathogenic bacteria for humans and Warm-blooded animals.

Thermophiles

- These grow best at temperatures above 45°C.

- The growth range of many thermopiles extends into the mesophilic region; these species are designated "Facultative thermophiles".

Oxygen Toxicity

- Oxygen is both beneficial and poisonous to living organisms.

- It is beneficial its strong oxidizing ability makes it an excellent terminal electron acceptor for the energy-yielding process known as "Respiration".

- Oxygen is also a toxic substance.

The following factors are among those that have been implicated in oxygen toxicity.

Oxygen Inactivation of Enzymes

Molecular oxygen can directly oxidize certain essential reduced groups such as thiol (-SH) groups (or) enzymes, results in enzyme inactivation.

Eg: Nitrogenase (nitrogen fixation enzymes) destroyed by even small amounts of oxygen.

Damage Due to Toxic Derivatives of Oxygen

Various cellular enzymes catalyze chemical reactions involving molecular oxygen; these reactions can result in the addition of a single electron to an O_2 molecule, by the result "Superoxide radical"(O_2^-).

$$O_2 + e^- \rightarrow O_{2-}$$

These superoxide radicals can inactivated vital cell components. Recent studies suggest the production of even more toxic substances by detrimental action of superoxide radical & protons form "Hydrogen peroxide (H_2O_2)" and "Hydroxyl radicals(OH.)".

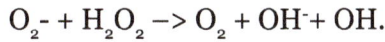

$$2\ O_2^- + 2H^+ \rightarrow O_2 + H_2O_2$$

$$O_2^- + H_2O_2 \rightarrow O_2 + OH^- + OH.$$

This superoxide radical will eliminate by greatly increasing the enzyme "Superoxide dismutase". These hydrogen peroxide produced by this reaction can, in turn, be dissipated by "Catalase" and "Peroxidase" enzymes.

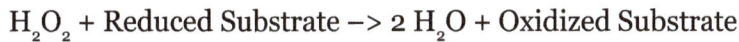

$$2\ H_2O_2 \rightarrow 2\ H_2O + O_2$$

$$H_2O_2 + \text{Reduced Substrate} \rightarrow 2\ H_2O + \text{Oxidized Substrate}$$

Elimination of either super-oxide radicals (or) hydrogen peroxide can prevent the formation of the highly dangerous hydroxyl radicals.

Acidity (or) Alkalinity (pH)

For most Bacteria, the optimum pH for growth lies between 6.5 and 7.5. Few Bacteria prefer more extreme pH values for growth. For example, "Thiobacillus thiooxidants" has optimum pH of 2.0 to 3.5 and can grow in a range between pH 0.5 and 6.0.

When Bacteria are cultivated in a medium originally adjusted to a given pH, for example, 7.0, it is very likely that this pH will change as a result of the chemical activities of the organisms.

Incorporating a buffer into the medium can prevent a radical shift in pH. A buffer is a mixture of a weak acid and its conjugate base (e.g.: Acetic acid (CH_3COOH) and acetate (CH_3COO^-). Phosphate buffer (combination of $H_2PO_4^-$ and $H\ PO4^{-2}$) having a pKa of 6.8, is widely used in bacteriological media.

Some large fermentation apparatus are equipped with automatic controls that maintain a constant pH.

Types of Bacterial Growth

Bacterial growth refers to the increase in cell number or cell population of bacteria. It is brought about by cell multiplication.

Methods of Bacterial Growth

Bacteria grow by the following methods of reproduction:

1. Binary fission   2. Budding

3. Filamentation   4. Sporulation

Binary Fission

Binary fission refers to the division of the parent cell into two daughter cells. During binary fission, both cytoplasm and nuclei divide.

E.g.: Bacillus, streptococcus, etc.

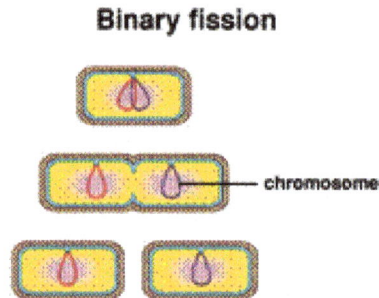

Binary fission in a). Bacillus b). Prosthecobacter.

Budding

In budding, the bacterial cell produces a small projection on its surface. It is called bud. It increases in size and separates into a daughter cell.

E.g.: Rhodopseudomonas

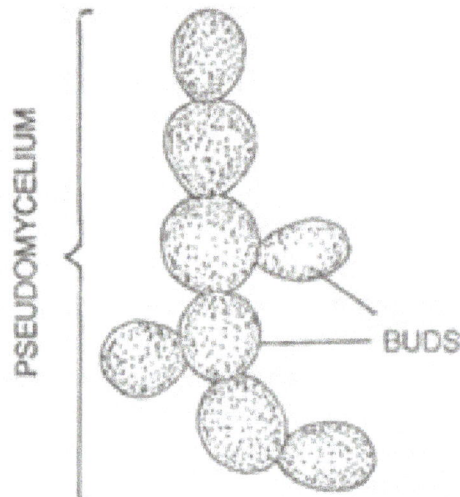

Filamentation

Filamentous Bacteria, the filaments break into fragments and each fragment grows into a daughter filament. This process of growth is called filamentation or fragmentation. E.g.: Nacardia.

Sporulation

Certain Bacteria grow by producing spores. The hypha at tip produces many spores. The spores separate and develop into new colonies.

E.g.: Streptomyces

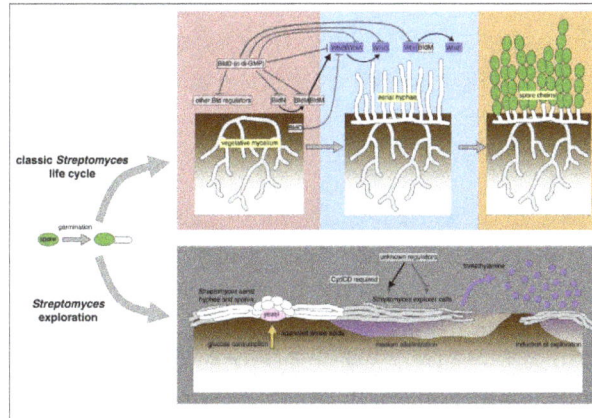

Bacterial Growth Rate

During growth, a cell divides into two and the two then divide into four unto 8 soon. In this way, the cells grow by geometrical progression.

$$1 \rightarrow 2 \rightarrow 4 \rightarrow 8 \rightarrow 16 \rightarrow 32 \rightarrow 64 \rightarrow$$

The time required by a cell to divide or the time required for the population to double is called generation time. The generation time varies from species to species and nutrients, temperature, etc affect it. The generation time for E.coli in milk at $37^{\circ}C$ is 12 minutes.

Number of cells (N) derived from a single cell after 'n' generations will be,

$$N = 2^n$$

Similarly, after 'n' generations, the number of cells derived from an initial cell population of NO will be,

$$N = 2^n \, NO$$

The rate of growth of a culture,

Different parameters are used to express the rate of growth in bacterial culture. They are:

Growth rate constant denoted as 'k', and

Mean doubling time or generation time denoted as 'g'.

The growth rate of a cell refers to how rapidly a cell increases in mass. The rate of increase in bacteria at any particular time is proportional to the number or mass of Bacteria present at that time. The bacterial growth equation is:

The Rate Increase of Cells = k (No. of Cells or Mass of Cells)

Here 'k' is the proportionality constant and is an index of the rate of growth; 'k' also to the amount of that cellular component. In mathematical terms,

$$\frac{dN}{dt} = kN; \quad \frac{dX}{dt} = kX; \quad and \quad \frac{dZ}{dt} = kZ$$

Here N is the number of cells per ml, X is the mass of cells/ml, Z is the amount of any cellular component/ml, 't' is the time and 'K' is the growth rate constant.

Generation Time (or) Mean Doubling Time

The generation time is defined as the time required for cell components of the culture or the number (or) mass to increase by a factor of 2. It is denoted by the symbol 'g'. Generation time is also referred to the time that a bacterial cell takes to go through its life cycle.

Generation time is usually shorter for prokaryotes than eukaryotes, a shorter for smaller than for larger cells since the growth rate is proportional to the energy metabolism of the cell. The faster the cell metabolizes nutrients, the shorter it's generation time.

Growth Curve

The increase in the cell size and cell mass during the development of an organism is termed as growth. It is the unique characteristics of all organisms. The organism must require certain basic parameters for their energy generation and cellular biosynthesis. The growth of the organism is affected by both physical and Nutritional factors. The physical factors include the pH, temperature, Osmotic pressure, Hydrostatic pressure, and Moisture content of the medium in which the organism is growing. The nutritional factors include the amount of Carbon, nitrogen, Sulphur, phosphorous, and other trace elements provided in the growth medium. Bacteria are unicellular (single cell) organisms. When the bacteria reach a certain size, they divide by binary fission, in which the one cell divides into two, two into four and continue the process in a geometric fashion. The bacterium is then known to be in an actively growing phase. To study the bacterial growth population, the viable cells of the bacterium should be inoculated on to the sterile broth and incubated under optimal growth conditions. The bacterium starts utilising the components of the media and it will increase in its size and cellular mass. The dynamics of the bacterial growth can be studied by plotting the cell growth (absorbance) versus the incubation time or log of cell number versus time. The curve thus obtained is a sigmoid curve and is known as a standard growth curve. The increase in the cell mass of the organism is measured by using the Spectrophotometer. The Spectrophotometer measures the turbidity or Optical density which is the measure of the amount of light absorbed by a bacterial suspension. The degree of turbidity in the broth culture is directly related to the number of microorganism present, either viable or dead cells, and is a convenient and rapid method of measuring cell growth rate of an organism. Thus the increasing the turbidity of the broth medium indicates increase of the microbial cell mass. The amount of transmitted light through turbid broth decreases with subsequent increase in the absorbance value.

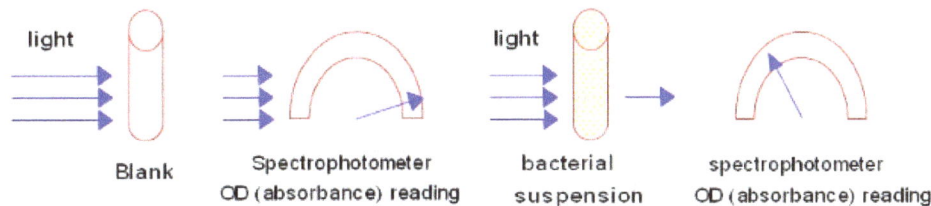

Figure: Absorbance reading of bacterial suspension.

The growth curve has four distinct phases.

1. Lag phase

When a microorganism is introduced into the fresh medium, it takes some time to adjust with the new environment. This phase is termed as Lag phase, in which cellular metabolism is accelerated, cells are increasing in size, but the bacteria are not able to replicate and therefore no increase in cell mass. The length of the lag phase depends directly on the previous growth condition of the organism. When the microorganism growing in a rich medium is inoculated into nutritionally poor medium, the organism will take more time to adapt with the new environment. The organism will start synthesising the necessary proteins, co-enzymes and vitamins needed for their growth and hence there will be a subsequent increase in the lag phase. Similarly when an organism from a nutritionally poor medium is added to a nutritionally rich medium, the organism can easily adapt to the environment, it can start the cell division without any delay, and therefore will have less lag phase it may be absent.

2. Exponential or Logarithmic (log) phase

During this phase, the microorganisms are in a rapidly growing and dividing state. Their metabolic activity increases and the organism begin the DNA replication by binary fission at a constant rate. The growth medium is exploited at the maximal rate, the culture reaches the maximum growth rate and the number of bacteria increases logarithmically (exponentially) and finally the single cell divide into two, which replicate into four, eight, sixteen, thirty two and so on (That is 2^0, 2^1, 2^2, 2^3.........2^n, n is the number of generations) This will result in a balanced growth. The time taken by the bacteria to double in number during a specified time period is known as the generation time. The generation time tends to vary with different organisms. E.coli divides in every 20 minutes, hence its generation time is 20 minutes, and for Staphylococcus aureus it is 30 minutes.

3. Stationary phase

As the bacterial population continues to grow, all the nutrients in the growth medium are used up by the microorganism for their rapid multiplication. This result in the accumulation of waste materials, toxic metabolites and inhibitory compounds such as antibiotics in the medium. This shifts the conditions of the medium such as pH and temperature, thereby creating an unfavourable environment for the bacterial growth. The reproduction rate will slow down, the cells undergoing division is equal to the number of cell death, and finally bacterium stops its division completely. The cell number is not increased and thus the growth rate is stabilised. If a cell taken from the stationary phase is introduced into a fresh medium, the cell can easily move on the exponential phase and is able to perform its metabolic activities as usual.

4. Decline or Death phase

The depletion of nutrients and the subsequent accumulation of metabolic waste products and other toxic materials in the media will facilitates the bacterium to move on to the Death phase. During this, the bacterium completely loses its ability to reproduce. Individual bacteria begin to die due to the unfavourable conditions and the death is rapid and at uniform rate. The number of dead cells exceeds the number of live cells. Some organisms which can resist this condition can survive in the environment by producing endospores.

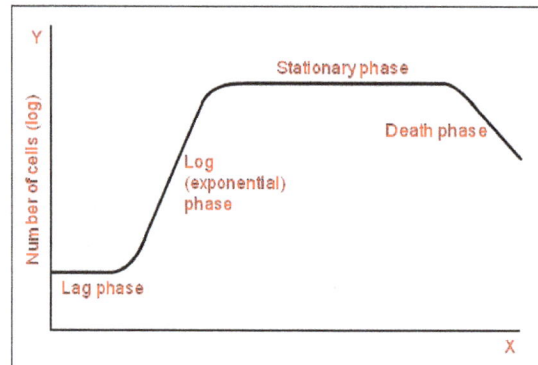

Figure: Different phases of growth of a bacteria.

Calculation

The generation time can be calculated from the growth curve.

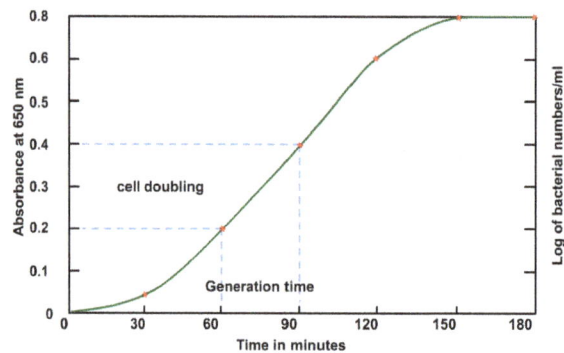

Figure: Calculation of generation time.

The exactly doubled points from the absorbance readings were taken and, the points were extrapolated to meet the respective time axis.

Generation Time =   (Time in minutes to obtain the absorbance 0.4) − (Time in minutes to obtain the absorbance 0.2)

> = 90-60

> = 30 minutes

Let N_0 = the initial population number

Nt = population at time t

N = the number of generations in time t

Therefore,

$$Nt = No \times 2^n$$

$$\log Nt = \log No + n \log 2$$

Therefore,

$$n = (\log Nt - \log No) / \log 2$$

$$n = (\log Nt - \log No) / 0.301$$

The growth rate can be expressed in terms of mean growth rate constant (k), the number of generations per unit time.

$$k = n / t$$

$$k = (\log Nt - \log No) / (0.301 \times t)$$

Mean generation time or mean doubling time (g), is the time taken to double its size.

Therefore,

$$Nt = 2No$$

Substituting equation $Nt = 2No$ in equation $k = (\log Nt - \log No) / (0.301 \times t)$

$$k = (\log Nt - \log No) / (0.301 \times t)$$

$$= (\log 2No - \log No) / (0.301 \times t)$$

$$= \log 2 + (\log No - \log No) / 0.301 \, g \quad \text{(Since the population doubles t= g)}$$

Therefore,

$$k = 1 / g$$

Mean growth rate constant, $k = 1 / g$

Mean generation time, $g = 1 / k$

Measuring Bacterial Growth

Estimating the number of bacterial cells in a sample, known as a bacterial count, is a common task performed by microbiologists. The number of bacteria in a clinical sample serves as an indication of the extent of an infection. Quality control of drinking water, food, medication, and even cosmetics relies on estimates of bacterial counts to detect contamination and prevent the spread of disease. Two major approaches are used to measure cell number. The direct methods involve counting cells, whereas the indirect methods depend on the measurement of cell presence or activity without actually counting individual cells. Both direct and indirect methods have advantages and disadvantages for specific applications.

Direct Cell Count

Direct cell count refers to counting the cells in a liquid culture or colonies on a plate. It is a direct way of estimating how many organisms are present in a sample. Let's look first at a simple and fast method that requires only a specialized slide and a compound microscope.

The simplest way to count bacteria is called the direct microscopic cell count, which involves transferring a known volume of a culture to a calibrated slide and counting the cells under a light microscope. The calibrated slide is called a Petroff-Hausser chamber and is similar to a hemocytometer used to count red blood cells. The central area of the counting chamber is etched into squares of various sizes. A sample of the culture suspension is added to the chamber under a coverslip that is placed at a specific height from the surface of the grid. It is possible to estimate the concentration of cells in the original sample by counting individual cells in a number of squares and determining the volume of the sample observed. The area of the squares and the height at which the coverslip is positioned are specified for the chamber. The concentration must be corrected for dilution if the sample was diluted before enumeration.

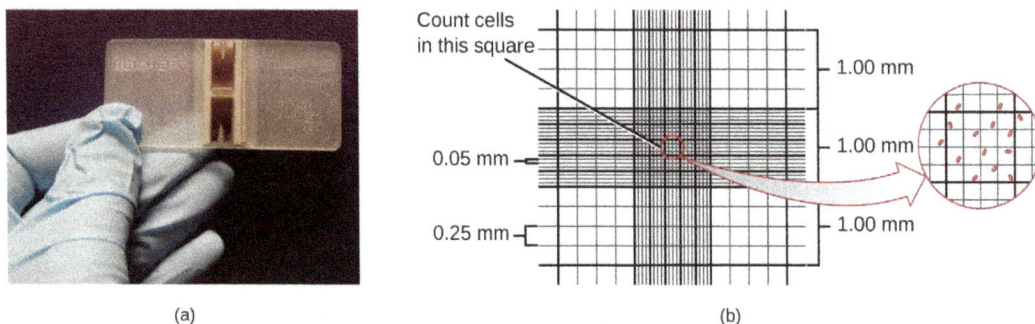

(a)   (b)

Figure: (a) A Petroff-Hausser chamber is a special slide designed for counting the bacterial cells in a measured volume of a sample. A grid is etched on the slide to facilitate precision in counting. (b) This diagram illustrates the grid of a Petroff-Hausser chamber, which is made up of squares of known areas. The enlarged view shows the square within which bacteria (red cells) are counted. If the coverslip is 0.2 mm above the grid and the square has an area of 0.04 mm2, then the volume is 0.008 mm3, or 0.000008 mL. Since there are 10 cells inside the square, the density of bacteria is 10 cells/0.000008 mL, which equates to 1,250,000 cells/mL.

Cells in several small squares must be counted and the average taken to obtain a reliable measurement. The advantages of the chamber are that the method is easy to use, relatively fast, and inexpensive. On the downside, the counting chamber does not work well with dilute cultures because there may not be enough cells to count.

Using a counting chamber does not necessarily yield an accurate count of the number of live cells because it is not always possible to distinguish between live cells, dead cells, and debris of the same size under the microscope. However, newly developed fluorescence staining techniques make it possible to distinguish viable and dead bacteria. These viability stains (or live stains) bind to nucleic acids, but the primary and secondary stains differ in their ability to cross the cytoplasmic membrane. The primary stain, which fluoresces green, can penetrate intact cytoplasmic membranes, staining both live and dead cells. The secondary stain, which fluoresces red, can stain a cell only if the cytoplasmic membrane is considerably damaged. Thus, live cells fluoresce green because they only absorb the green stain, whereas dead cells appear red because the red stain displaces the green stain on their nucleic acids.

Another technique uses an electronic cell counting device (Coulter counter) to detect and count the changes in electrical resistance in a saline solution. A glass tube with a small opening is immersed in an electrolyte solution. A first electrode is suspended in the glass tube. A second electrode is located outside of the tube. As cells are drawn through the small aperture in the glass tube, they briefly change the resistance measured between the two electrodes and the change is recorded by an electronic sensor; each resistance change represents a cell. The method is rapid and accurate within a range of concentrations; however, if the culture is too concentrated, more than one cell may pass through the aperture at any given time and skew the results. This method also does not differentiate between live and dead cells.

Direct counts provide an estimate of the total number of cells in a sample. However, in many situations, it is important to know the number of live, or viable, cells. Counts of live cells are needed when assessing the extent of an infection, the effectiveness of antimicrobial compounds and medication, or contamination of food and water.

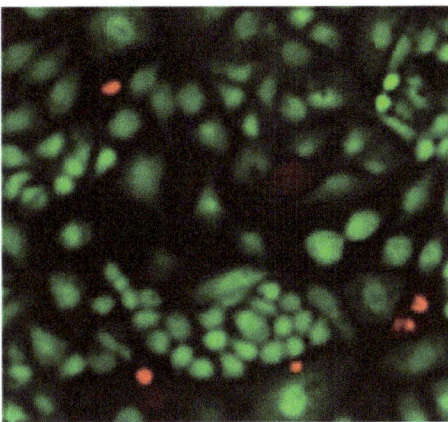

Figure: Fluorescence staining can be used to differentiate between viable and dead bacterial cells in a sample for purposes of counting. Viable cells are stained green, whereas dead cells are stained red.

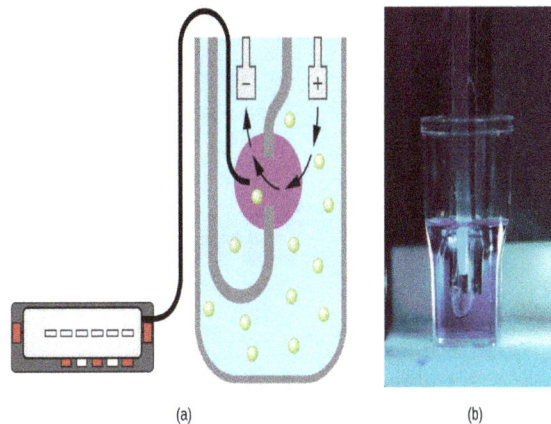

Figure: A Coulter counter is an electronic device that counts cells. It measures the change in resistance in an electrolyte solution that takes place when a cell passes through a small opening in the inside container wall. A detector automatically counts the number of cells passing through the opening.

Plate Count

The viable plate count, or simply plate count, is a count of viable or live cells. It is based on the principle that viable cells replicate and give rise to visible colonies when incubated under suitable conditions for the specimen. The results are usually expressed as colony-forming units per milliliter (CFU/mL) rather than cells per milliliter because more than one cell may have landed on the same spot to give rise to a single colony. Furthermore, samples of bacteria that grow in clusters or chains are difficult to disperse and a single colony may represent several cells. Some cells are described as viable but nonculturable and will not form colonies on solid media. For all these reasons, the viable plate count is considered a low estimate of the actual number of live cells. These limitations do not detract from the usefulness of the method, which provides estimates of live bacterial numbers.

Microbiologists typically count plates with 30–300 colonies. Samples with too few colonies (<30) do not give statistically reliable numbers, and overcrowded plates (>300 colonies) make it difficult to accurately count individual colonies. Also, counts in this range minimize occurrences of more than one bacterial cell forming a single colony. Thus, the calculated CFU is closer to the true number of live bacteria in the population.

There are two common approaches to inoculating plates for viable counts: the pour plate and the spread plate methods. Although the final inoculation procedure differs between these two methods, they both start with a serial dilution of the culture.

Serial Dilution

The serial dilution of a culture is an important first step before proceeding to either the pour plate or spread plate method. The goal of the serial dilution process is to obtain plates with CFUs in the range of 30–300, and the process usually involves several dilutions in multiples of 10 to simplify calculation. The number of serial dilutions is chosen according to a preliminary estimate of the culture density. Figure 10 illustrates the serial dilution method.

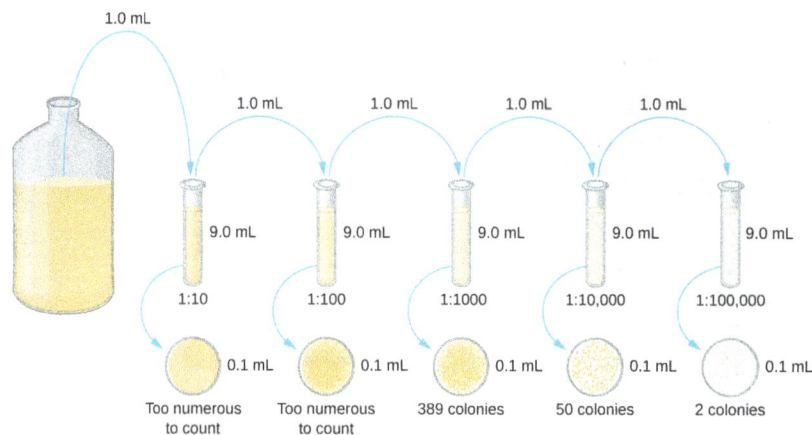

Figure: Serial dilution involves diluting a fixed volume of cells mixed with dilution solution using the previous dilution as an inoculum. The result is dilution of the original culture by an exponentially growing factor.

A fixed volume of the original culture, 1.0 mL, is added to and thoroughly mixed with the first dilution tube solution, which contains 9.0 mL of sterile broth. This step represents a dilution factor of 10, or 1:10, compared with the original culture. From this first dilution, the same volume, 1.0 mL, is withdrawn and mixed with a fresh tube of 9.0 mL of dilution solution. The dilution factor is now 1:100 compared with the original culture. This process continues until a series of dilutions is produced that will bracket the desired cell concentration for accurate counting. From each tube, a sample is plated on solid medium using either the pour plate method or the spread plate method. The plates are incubated until colonies appear. Two to three plates are usually prepared from each dilution and the numbers of colonies counted on each plate are averaged. In all cases, thorough mixing of samples with the dilution medium (to ensure the cell distribution in the tube is random) is paramount to obtaining reliable results.

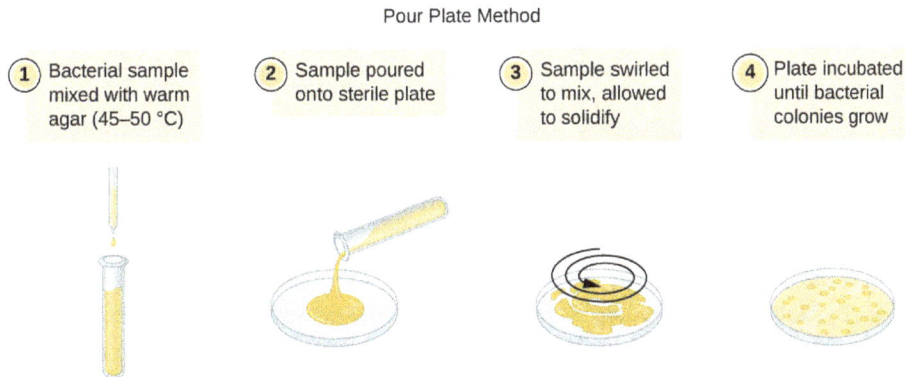

Pour Plate Method

1. Bacterial sample mixed with warm agar (45–50 °C)
2. Sample poured onto sterile plate
3. Sample swirled to mix, allowed to solidify
4. Plate incubated until bacterial colonies grow

Figure: In the pour plate method of cell counting, the sample is mixed in liquid warm agar (45–50°C) poured into a sterile Petri dish and further mixed by swirling. This process is repeated for each serial dilution prepared. The resulting colonies are counted and provide an estimate of the number of cells in the original volume sampled.

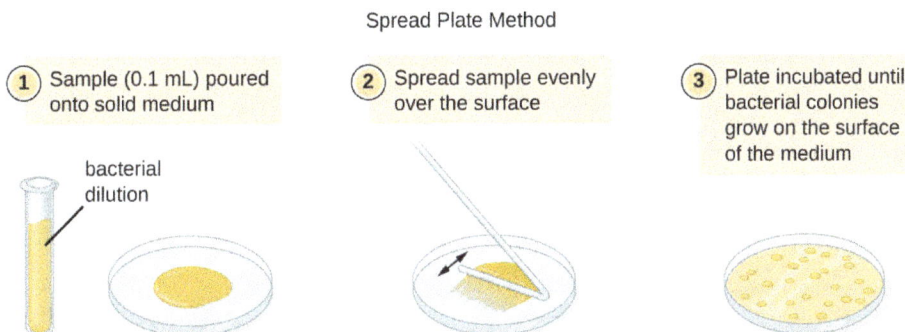

Spread Plate Method

1. Sample (0.1 mL) poured onto solid medium
2. Spread sample evenly over the surface
3. Plate incubated until bacterial colonies grow on the surface of the medium

bacterial dilution

Figure: In the spread plate method of cell counting, the sample is poured onto solid agar and then spread using a sterile spreader. This process is repeated for each serial dilution prepared. The resulting colonies are counted and provide an estimate of the number of cells in the original volume samples.

The dilution factor is used to calculate the number of cells in the original cell culture. In our example, an average of 50 colonies was counted on the plates obtained from the 1:10,000 dilution. Because only 0.1 mL of suspension was pipetted on the plate, the multiplier required to reconstitute the original concentration is 10 × 10,000. The number of CFU per mL is equal to 50 × 100 × 10,000 = 5,000,000. The number of bacteria in the culture is estimated as 5 million cells/mL. The colony count obtained from the 1:1000 dilution was 389, well below the expected 500 for a 10-fold difference in dilutions. This highlights the issue of inaccuracy when colony counts are greater than 300 and more than one bacterial cell grows into a single colony.

A very dilute sample—drinking water, for example—may not contain enough organisms to use either of the plate count methods described. In such cases, the original sample must be concentrated rather than diluted before plating. This can be accomplished using a modification of the plate count technique called the membrane filtration technique. Known volumes are vacuum-filtered aseptically through a membrane with a pore size small enough to trap microorganisms. The membrane is transferred to a Petri plate containing an appropriate growth medium. Colonies are counted after incubation. Calculation of the cell density is made by dividing the cell count by the volume of filtered liquid.

Most Probable Number

The number of microorganisms in dilute samples is usually too low to be detected by the plate count methods described thus far. For these specimens, microbiologists routinely use the most probable number (MPN) method, a statistical procedure for estimating of the number of viable microorganisms in a sample. Often used for water and food samples, the MPN method evaluates detectable growth by observing changes in turbidity or color due to metabolic activity.

A typical application of MPN method is the estimation of the number of coliforms in a sample of pond water. Coliforms are gram-negative rod bacteria that ferment lactose. The presence of coliforms in water is considered a sign of contamination by fecal matter. For the method illustrated in figure, a series of three dilutions of the water sample is tested by inoculating five lactose broth tubes with 10 mL of sample, five lactose broth tubes with 1 mL of sample, and five lactose broth tubes with 0.1 mL of sample. The lactose broth tubes contain a pH indicator that changes color from red to yellow when the lactose is fermented. After inoculation and incubation, the tubes are examined for an indication of coliform growth by a color change in media from red to yellow. The first set of tubes (10-mL sample) showed growth in all the tubes; the second set of tubes (1 mL) showed growth in two tubes out of five; in the third set of tubes, no growth is observed in any of the tubes (0.1-mL dilution). The numbers 5, 2, and 0 are compared with Mathematical Basics figure, which has been constructed using a probability model of the sampling procedure. From our reading of the table, we conclude that 49 is the most probable number of bacteria per 100 mL of pond water.

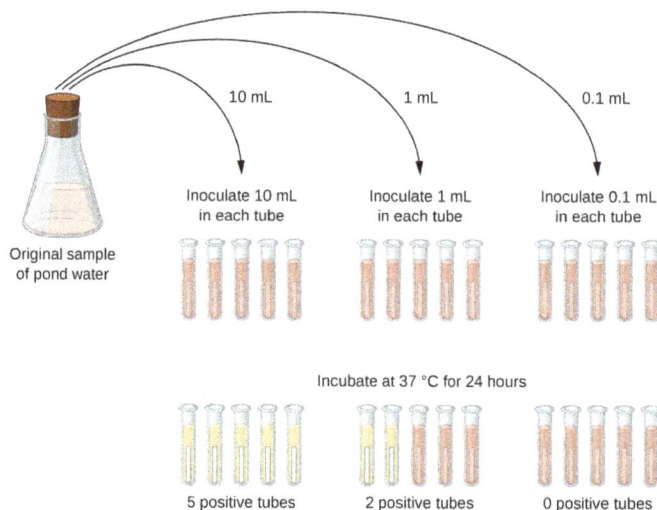

Figure: In the most probable number method, sets of five lactose broth tubes are inoculated with three different volumes of pond water: 10 mL, 1 mL, and 0.1mL. Bacterial growth is assessed through a change in the color of the broth from red to yellow as lactose is fermented.

Indirect Cell Counts

Besides direct methods of counting cells, other methods, based on an indirect detection of cell density, are commonly used to estimate and compare cell densities in a culture. The foremost approach is to measure the turbidity (cloudiness) of a sample of bacteria in a liquid suspension. The laboratory instrument used to measure turbidity is called a spectrophotometer. In a spectrophotometer, a light beam is transmitted through a bacterial suspension, the light passing through the suspension is measured by a detector, and the amount of light passing through the sample and reaching the detector is converted to either percent transmission or a logarithmic value called absorbance (optical density). As the numbers of bacteria in a suspension increase, the turbidity also increases and causes less light to reach the detector. The decrease in light passing through the sample and reaching the detector is associated with a decrease in percent transmission and increase in absorbance measured by the spectrophotometer.

Measuring turbidity is a fast method to estimate cell density as long as there are enough cells in a sample to produce turbidity. It is possible to correlate turbidity readings to the actual number of cells by performing a viable plate count of samples taken from cultures having a range of absorbance values. Using these values, a calibration curve is generated by plotting turbidity as a function of cell density. Once the calibration curve has been produced, it can be used to estimate cell counts for all samples obtained or cultured under similar conditions and with densities within the range of values used to construct the curve.

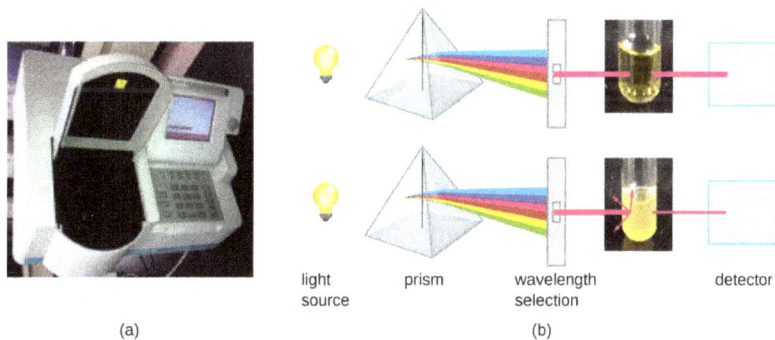

light source   prism   wavelength selection   detector

(a)   (b)

Figure: (a) A spectrophotometer is commonly used to measure the turbidity of a bacterial cell suspension as an indirect measure of cell density. (b) A spectrophotometer works by splitting white light from a source into a spectrum. The spectrophotometer allows choice of the wavelength of light to use for the measurement. The optical density (turbidity) of the sample will depend on the wavelength, so once one wavelength is chosen, it must be used consistently. The filtered light passes through the sample (or a control with only medium) and the light intensity is measured by a detector. The light passing into a suspension of bacteria is scattered by the cells in such a way that some fraction of it never reaches the detector. This scattering happens to a far lesser degree in the control tube with only the medium.

Measuring dry weight of a culture sample is another indirect method of evaluating culture density without directly measuring cell counts. The cell suspension used for weighing must be concentrated by filtration or centrifugation, washed, and then dried before the measurements are taken. The degree of drying must be standardized to account for residual water content. This method is especially useful for filamentous microorganisms, which are difficult to enumerate by direct or viable plate count.

As we have seen, methods to estimate viable cell numbers can be labor intensive and take time because cells must be grown. Recently, indirect ways of measuring live cells have been developed that are both fast and easy to implement. These methods measure cell activity by following the production of metabolic products or disappearance of reactants. Adenosine triphosphate (ATP) formation, biosynthesis of proteins and nucleic acids, and consumption of oxygen can all be monitored to estimate the number of cells.

Biofilms

Biofilm is an association of micro-organisms in which microbial cells adhere to each other on a living or non-living surfaces within a self-produced matrix of extracellular polymeric substance. Bacterial biofilm is infectious in nature and can results in nosocomial infections. According to National Institutes of Health (NIH) about about 65% of all microbial infections, and 80% of all chronic infections are associated with biofilms. Biofilm formation is a multi-step process starting with attachment to a surface then formation of micro-colony that leads to the formation of three dimensional structure and finally ending with maturation followed by detachment. During biofilm formation many species of bacteria are able to communicate with one another through specific mechanism called quorum sensing. It is a system of stimulus to co-ordinate different gene expression. Bacterial biofilm is less accessible to antibiotics and human immune system and thus poses a big threat to public health because of its involvement in variety of infectious diseases. A greater understanding of bacterial biofilm is required for the development of novel, effective control strategies thus resulting improvement in patient management.

Composition of Biofilms

Biofilms are group or micro-organisms in which microbes produced an extracellular polymeric substances (EPS) such as proteins (<1-2%) including enzymes), DNA (<1%), polysaccharides (1-2%) and RNA (<1%), and in addition to these components, water (up to 97%) is the major part of biofilm which is responsible for the flow of nutrients inside biofilm matrix. The architecture of biofilm consists of two main components i.e. water channel for nutrients transport and a region of densely packed cells having no prominent pores in it. The microbial cells with in biofilms are arranged in way with significant different physiology and physical properties. Bacterial biofilms are normally beyond the access of antibiotics and human immune system. Microorganisms that produce biofilm have enhanced potential to bear and neutralize antimicrobial agents and result in prolonged treatment. Biofilm forming bacteria switch on some genes that activate the expression of stress genes which in turn switch to resistant phenotypes due certain changes e.g. cell density, nutritional or temperature, cell density, pH and osmolarity. When the biofilm water channels are compared with system of circulations showed that biofilms are considered primitive multi-cellular organisms. Various components of biofilms are shown in table below signify the biofilm integrity and making it resistant against various environmental factors.

S. No	Components	Percentage of matrix
1	Microbial cells	2-5%
2	DNA and RNA	<1-2%
3	Polysaccharides	1-2%
4	Proteins	<1-2% (including enzymes)
5	Water	Up to 97%

Table: Biofilm chemical composition

Formation of Biofilms

Biofilm formation is a highly complex process, in which microorganism cells transform from planktonic to sessile mode of growth. It has also been suggested that biofilm formation is dependent on the expression of specific genes that guide the establishment of biofilm. The process of biofilm formation occurs through a series of events leading to adaptation under diverse nutritional and environmental conditions. This is a multi-step process in which the microorganisms undergo certain changes after adhering to a surface. Microorganisms which form biofilms are shown to elicit specific mechanisms. Biofilm formation has following important steps (a) attachment initially to a surface (b) formation of micro-colony (c) three dimensional structure formation (d) biofilm formation, maturation and detachment (dispersal).

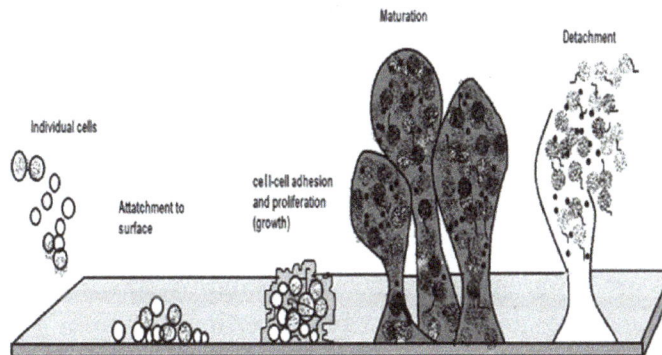

Figure: the biofilm life cycle in three steps: attachment, growth of colonies
(micro-colony formation and formation of three dimensional structures) and detachment in clumps.

Attachment

When a bacterium cell reaches to near some surface/support so close that its motion is very slow down, it make a reversible connection with the surface and/or already adhered other microbe to the surface. For biofilm formation, a system of solid– liquid interface can provide an ideal environment for micro-organism to attach and grow (e.g. blood, water). For most frequent attachment and biofilm formation rough, hydrophilic and coated surfaces will provide better environment. Increased attachment may also occur due to increase but not exceeding critical level in flow velocity, temperature of water or nutrients concentrations. Presence of locomotor structures on cell surfaces such as flagella, pili, fimbriae, proteins or polysaccharides are also important and may possibly provide an advantage in biofilm formation when there are mixed community.

Micro-colony Formation

Micro-colony formation takes place after bacteria adhered to the physical surface/biological tissue and this binding then becomes stable which results in formation of micro-colony. Multiplication of bacteria in the biofilm starts as a result of chemical signals. The genetic mechanism of exopolysaccharide production is activated when intensity of the signal cross certain threshold. So by this way using such chemical signal, the bacterial cell divisions take place within the embedded exopolysaccharide matrix, which finally result in micro-colony formation.

Three-dimensional Structure Formation and Maturation

After micro-colony formation stage of biofilm, expression of certain biofilm related genes take place. These gene products are needed for the EPS which is the main structure material of biofilm. It is reported that bacterial attachment by itself can trigger formation of extracellular matrix.

Matrix formation is followed by water-filled channels formation for transport of nutrients within the biofilm. Researcher have proposed that these water channels are like a circulatory systems, distributing different nutrients to and removing waste materials from the communities in the micro-colonies of the biofilm.

Detachment

After biofilm formation, the researchers have often noticed that bacteria leave the biofilms itself on regular basis. By doing this the bacteria can undergo rapid multiplication and dispersal. Detachment of planktonic bacterial cells from the biofilm is a programmed detachment, having a natural pattern . Sometime occasionally due to some mechanical stress bacteria are detached from the colony into the surrounding. But in most cases some bacteria stop EPS production and are detached into environment. Dispersing of biofilm cells occur either by detachment of new formed cells from growing cells or dispersion of biofilm aggregates due to flowing effects or due to quorum- sensing . In biofilm of cells are removed due to an enzyme action that causes alginate digestion . Phenotypic characters of organisms are apparently affected by the mode of biofilm dispersion. Dispersed cells from the biofilm have the ability to retain certain properties of biofilm, such as antibiotic in-sensitivity. The cells which are dispersed form biofilm as result of growth may return quickly to their normal planktonic phenotype. The different steps in biofilm life cycle are shown in figure below.

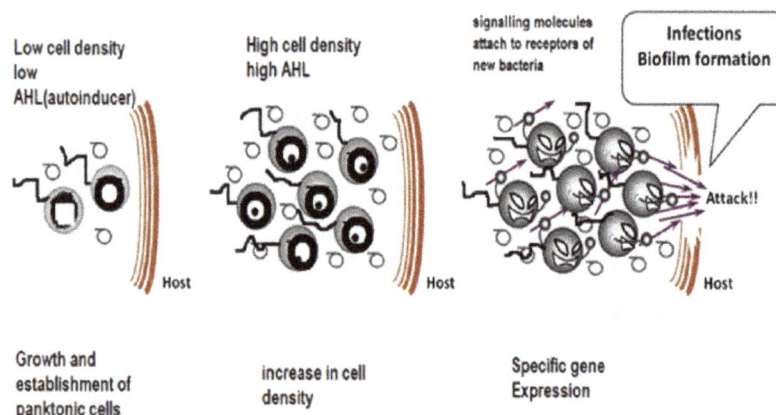

Figure: Cell density dependent gene expression in quorem sensing.

Quorum Sensing

During biofilm formation many species of bacteria are able to communicate with one another through a mechanism called quoreum sensing. It is a system of stimulus to co-ordinate gene expression with other cells and response related to the density of their local population. During quorum sensing signalling molecules attach to receptors of new bacteria and help in transcription of genes within a single species of bacteria as well as between different bacterial species. QS system enables communication between intraspecies and interspecies which involves in terms of biofilm formation, food shortages and environmental stress conditions, such as disinfectants, antibiotics, bacterial colonization, the identification of annoying species, the establishment of normal intestinal flora as well as the prevention of harmful intestinal flora. Many clinically-associated bacteria use QS for the regulation of the collective production of virulence factors. QS in Gram-positive bacteria occur through a series of events such as production, detection and response to AIs. The oligopeptide auto inducing peptides in many Gram-positive bacteria are detected by membrane-bound two component signal transduction systems. They are encoded as precursors (pro AIPs) and possess sequence diversity.

Biofilm Forming Bacteria

Nearly all (99.9%) of micro-organisms have the ability to form biofilm on a wide range of surfaces i.e. biological and inert surfaces. When micro-organisms bind to a surface, they produce extracellular polymeric substance (EPS) and form biofilm. Biofilm posing a great problem for public health due to its resistant nature to antibiotics and disease associated with indwelling medical devices. It is found that H. influenza has the ability to form biofilm in human body and can escape from human immune system . Biofilm forming capability has been reported in large number of bacterial species such as P. aeruginosa, S. epidermidis, E. coli spp, S. aureus, E. cloacae, K. pneumoniae. A few biofilm forming bacterial species have described below.

S.No	Common biofilm forming bacterial species
1	E. coli
2	P. aeruginosa,
3	S. epidermidis,
4	S. aureus,
5	Staphylococcus epidermidis
6	E. cloacae
7	K. pneumoniae
8	Actenomyces israelii
9	Haemophilus influenza
10	Burkholderia cepacia

Table: A list of common biofilm forming bacterial species.

Escherichia Coli

E. coli is a rod shaped Gram negative bacteria causing a large number of nosocomial and community infections such as urinary tract infections (UTIs) and prostatitis. It has the ability to secret toxins, polysaccharide and can form biofilm. It can also form biofilm in-vitro . E. coli capsules are high molecular weight molecules and are attached to the cell surface. E. coli capsule play an indirect role in biofilm by protecting bacterial surface adhesion. Different environmental conditions affect E. coli capability to form biofilm. Thickness of E. coli biofilm may be of hundreds of microns and posing a difficulty in treatment with antibiotics due to presence of exopolymers.

Pseudomonas Aeruginosa

P. aeruginosa is a Gram negative notorious opportunistic pathogen found along with other Pseudomonas species as part of normal flora of human skin. P. aeruginosa is a ubiquitous human pathogenic organism present everywhere, and can be isolated from different sources such as humans, plants and animals . It has a strong tendency to form biofilm and such biofilm has been found to be partially responsible for chronic infections. P. aeruginosa biofilm can be eradicated by using silver. Silver has antibacterial activity and bactericidal concentration of silver necessary to eradicate the bacterial biofilm was found to be 10-100 times higher as compared to that used to eradicate plank-tonic bacteria. The biofilms of P. aeruginosa are developed communities of individual cells that are encased in an extracellular polysaccharide matrix. These are extremely resistant to antibiotics. Their biofilm formation involves an initial attachment to a solid surface which leads to the forma-tion of micro-colonies. These micro colonies differentiate into exopolysaccharide-encased, mature biofilms. P. aeruginosa is a multidrur resistantant bacteria even including ciprofloxacin, that are commonly used for the treatment of lung infections .

Staphylococcus Aureus

S. aureus is a multi-drug resistant bacteria causing a number of nosocomial infections. It grows on catheters and chronic wounds as biofilm. S. aureus recycles proteins for the formation of the extracellular matrix in the cytoplasm. The cytoplasmic proteins also working as matrix proteins allows enhanced flexibility and adaptation to S. aureus in forming biofilms in infectious conditions and could encourage the formation of mixed-species biofilms in chronic wounds.

Stereptococcus Epidermidis

S. epidermidis is well known as an opportunistic pathogen that has greater potential to cause in-fections in patients with immune-compromised state, intravenous drug abusers, AIDS patients, immuno-suppressive therapy patients and premature new born . During surgical implantation of polymeric devices the S.epidermidis contamination chances increases because of its biofilm form-ing capability. Biofilm formation is responsible for device related infections of S. epidermidis and this consequently leads to the pathogenesis. In staphylococci, the main factor which is responsible for adhesion is called the polysaccharide intercellular adhesion (PIA). In Staphylococcus cells are covers by PIA which hold them together as the most important component of the extracellular matrix. In a recent study Rohde have reported PIA-independent biofilm formation in about 27% of biofilm-forming S. epidermidis strains isolated from prosthetic joint infections.

Enterobacter Cloacae

E. cloacae is a Gram positive bacteria causing a range of nosocomial infections in human i.e. lower respiratory tract infection, bacteraemia, urinary tract infections, endocarditis, intra-abdominal in-fections, septic arthritis, skin and soft tissue infections, osteomyelitis and ophthalmic infections. En-terobacter causing nasocomial infections is most frequently isolated species and in recent years has emerged as important pathogenic bacteria. Bloodstream infections which are responsible for morbid-ity and mortality in both developing and developed countries are caused by E. cloacae. It also causes biofilm associated infections such as UTIs and biliary tract infections. Enterobector also has property of intrinsic resistance to certain antibiotics such as ampicillin and narrow-spectrum cephalosporins. It is also showing high frequency of mutations to expanded-spectrum and broadspectrum cephalo-sporin. It also possesses ß-lactamase and showing resistance to third generation cephalosporins.

Klebsiella Pneumoniae

K. pneumoniae is a Gram negative bacterium, frequently causing nosocomial infections, belongs to the genus Klebsiella. K. pneumonie is very important species among genus Klebsiella and causing a considerable proportion of nosocomial infections such as urinary tract infections (UTI), pneumonia, septicemias and soft tissue infections . Field emission scanning electron microscopy (F-ESEM) and confocal laser scanning microscopy were used to investigate the biofilm of K. pneumoniae, isolated from clinical strains. In a study on different strains of K. pneumonia isolated from various human samples such as urine, blood sputum and wound swabs, it was reported that about 40% strains had the ability to form biofilms. However, increasing temperature from 35°C to 40°C resulted in consistent growth of biofilm on abiotic surfaces. Also the ability of Klebsiella pneumonia to form biofilm takes place more successfully in a mix strains than individual strain.

Biofilms and Antibiotic Resistance

Mechanisms of antibiotics and biocides resistance of biofilms are categorized into four classes which include (a) active molecule inactivation directly (b) altering body's sensitivity to target of action, (c) reduction of the drug concentration before reaching to the target site and (d) efflux systems. Biofilm antibiotic resistance level may vary among different sittings and the key factors responsible for this resistance may also differ. Regarding resistance, the primary evidence shows that conventional mechanisms are unable to explain the high resistance to antibacterial agents associated with biofilms, although this evidence cannot be ignored in resistance in the growth of adherent cells. So it is suggested that the resistance posed by the adhered bacteria or biofilms may have some intrinsic mechanisms and are responsible for conventional antibiotic resistance. Several mechanisms have been explored that are considered to be key factors in high resistance nature of biofilms. These mechanisms are (a) limited diffusion, (b) enzyme causing neutralizations, (c) heterogeneous functions, (d) slow growth rate, (e) presence of persistent (non-dividing) cells and (f) biofilm phenotype such adaptive mechanisms e.g. efflux pump and membrane alteration .

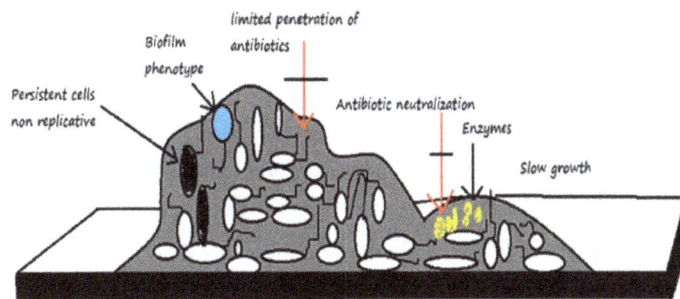

Figure: Antibiotic resistance associate to biofilm. Description of the key involved in antibiotic resistance such as enzyme causing neutralization, presence of persistent (non- dividing) cells and biofilm phenotype.

Low Penetration of Antibiotics

Diffusion of antibiotics can take place through the matrix of the biofilm. Diffusion or penetration of antibiotics to deeper layers of biofilm is affected by exopolysaccharide acting as a physical barrier. When molecules direct interact with this matrix, their movement to the interior of the biofilm is slow down, resulting antibiotic resistance. This may also acts as a hindrance for high molecular weight molecules such as complement system proteins and lysozyme, and in liquid culture bacterial cells are readily exposed to antibiotics as compare to compact structure biofilm. Bacteria escape from biofilm that do not produce polysaccharide and are easily attack by immune system cells.

Inactivation of antibiotic takes place when bind to biofilm matrix. P. aeruginosa have alginate exopolysaccharide, which is anionic in nature. Presence of this matrix explains slow penetration of fluoroquinolones and aminoglycosides. Low penetration of antibiotic is not sufficient to explain the biofilm resistance, other mechanisms have been assumed that must be involved. This is also suggested recently that slow diffusion of antibiotics permit plenty of time to establish a protective response to stress.

Neutralization by Enzymes

Antibiotics resistance in biofilm may be due to the presence of neutralizing enzymes which degrade or inactivate antibiotics. These enzymes are proteins which confer resistance by mechanisms such as hydrolysis, modification of antimicrobials by different biochemical reactions. Accumulations of these enzymes occur in the glycocalyx from the biofilm surface by the action of antibiotics. Neutralization by enzymes is enhanced by slow penetration of antibiotics and also antibiotics degradation in the biofilm. In cystic fibrosis which is caused by P. aeruginosa, overproduction of cephalosporinase AmpC enzymes is responsible for resistance to different antibiotics. This enzyme confers resistance to β-lactam in the presence of even much more concentration of carbapenems . During a study when filters impregnated with antibiotics was applied on K. pneumoniae biofilm (mutant cells β-lactamases), in spite of good diffusion of antibiotic, growth was observed, suggesting that there would be another mechanism of resistance which need to be explored.

Heterogeneous Nature

Studies performed on determination of microbial growth in biofilms by using a microelectrode with probes to direct measure oxygen concentration in different areas of the biofilms . The biofilms are heterogeneous nature both metabolically and structurally and both processes such as aerobic and anaerobic occur at the same time. So response against antibiotics may be different in different areas of the biofilms. On surface of biofilm there is a high level of activity of antibiotics while inside the biofilms, slow or absent growth reduces the sensitivity of the cells to antimicrobials. In the various sub layers of biofilm, aerobic or facultative anaerobic microbial populations help us to know the differential susceptibility to various antibiotic therapies. Antibiotics response to the planktonic forms is different from the adhered cells. Action of aminoglycosides is affected by limitation of oxygen and anaerobic growth of microorganisms, which is affected by the presence of oxygen and pH gradients.

Cells Slow Growth Rate

Slow growth of microorganisms occurs due to limited availability of nutrients which confer resistance to antibiotics. In case of biofilm a gradient of nutrients resulting in metabolically active cell (periphery or surface layer) and inactive cells (within its interior). Bacterial cells are attack by both penicillin and ampicillin only when they are growing. Some other antibiotics that attack cells in stationary phase are β-lactams, aminoglycosides, cephalosporin and fluoroquinolones . So due to slow growth resistance has been determined in different bacterial strains such as resistance to cetrimide on E. coli, piperacillin and tobramycin in P. aeruginosa and ciprofloxacin on S. epidermidis. It has been shown that this resistance was due to slow growth . There are some natural peptides produce during host innate immune response act as antibacterial providing protection to the body. A peptide named polymyxin E is effective in treating cancer patient's biofilms and cystic fibrosis, caused by P. aeruginosa . In Cystic fibrosis patients, using ciprofloxacin and tetracycline can clear active growing cells and it is suggested that a combination of antibiotic colistin with other two antibiotics (ciprofloxacin and tetracycline) will be very effective in clearing P. aeruginosa.

Existence of Persistent Cells

After a purling antibiotics treatment of biofilm, a very small number of bacterial cells remain viable, called persistent cells. These cells may or may not give this resistance to their progeny and return to their normal state after the release of the applied stress or pressure. The persistent cells stop their replication for small duration for the survival of the community. Their adaptive mechanism is not related to the mechanism followed by the cells during stress (environmental damage). Persistent cells can bear multiple antibiotic doses and work for survival. When density of bacterial cells number in stationary phase raised to maximum, persistent cells increase in number indicting their main role in survival. There are certain evidences for the presence of persistent cells in biofilm: a) there is existence of a biphasic dimension in biofilms which means that large number of cells population is attacked while the rest of the population is not attacked (resistant) even with an extensive antibiotics treatment, b) persistence gene description function as a circuits of regulation, c) bacteriostatic antibiotics contribute to the growth of persistent cell and biofilm preservation by inhibiting growth of sensitive cells and d) reshaping of biofilm into original form when the antibiotics therapy is withdrawn.

Biofilm Phenotype

During biofilm formation, bacteria produce some products called secondary metabolites. These products are not required by the cell for their growth. These metabolites function as signaling molecules thus enhancing formation process of biofilms. Biofilm phenotype is regarded as community cells that confer no response to antibiotics treatment. These characteristics have proposed the presence of specific genes. In B. subtilis only 6% difference in gene expression was observed for biofilm by DNA microarrays as compared to their planktonic culture cells, while for P. aeruginosa this difference was only 1%. However at present time, this differential gene expression has not been proven fruitful for describing this mechanism.

Efflux Pumps

Efflux pumps are protein structures, either express constitutively or intermittently. These pumps may have substrate specificity. Similar compounds can be transported by these pumps that may be involved in multidrug resistance. Efflux pumps, inside the bacteria in the periplasmic area, are involved in antagonized accumulation of antibiotics. The show resistant to multiple antibiotics such as tetracycline, macrolides, fluoroquinolones, β-lactam and thus reducing these antibiotics concentration at low toxic level. Five families of efflux transporters have been identified in prokaryotes. Over expression of the efflux pumps have been considered to be responsible for antibiotic resistance in P. aeruginosa biofilms.

Membrane Proteins Alterations

Permeability of outer membrane is very important for the antibiotic diffusion through different routes. A key role is played in transportation of hydrophilic molecules by outer membrane channel proteins (porins) present in Gram-negative bacteria from the outer environment to the periplasmic space. Mutation in porins encoding genes can result in production of non-functional or altered proteins. These mutant porins have low permeability for the passage of hydrophobic molecules . OprD is a specific porin present in P. aeruginosa which enhances the absorbance of basic amino acids and imipenem. Loss of OprD in P. aeruginosa is responsible for resistance to imipenem, resulting in three-dimensional disturbance of imipenem molecule. The differential expression of porins coding genes, occur in biofilm, leading to antibiotic resistance. When the expression of ompC and three other genes (osmotically regulated) is increased then the bacterial cells grow as biofilms.

Phase Variation

Diverse phenotype within biofilms plays a significant role and is responsible for the resistant infections. Authors have reported this phenomenon for many genera and species which are exemplified by Pseudomonas and Staphylococcus genus, and also some species of Enterobacteriaceae. Biofilms have the capability to develop bacterial subpopulation to switch to the quiescent state as small-colony known as small colony variants (SCVs). These variants cells have very less susceptibility to growth phase dependent killing of antibiotics. Also have a defective catalase activity which interferes with oxidative metabolism. Detectable colonial morphological changes caused by SCVs in biofilms, leads to increased adherence, auto-aggregation, increased hydrophobicity and low level motility. It can also withstand wide range of harsh environmental stress conditions so this is considered as survival mechanism for biofilms. Phase variation was considered as the process of cellular internal rearrangement but recently it is consider as phase variation that occur due to genetic elements interaction. Various mechanisms involved in antibiotic resistance due to biofilm.

Biofilm Role in Infections

Microscopic evaluation of specimens from chronic wounds often indicates the presence of biofilms. Traditionally, three phases are use to describe wound microbiology. These three phases described as: contamination, colonization, and infection. Contamination refers to the presence of bacteria in the wound, whereas the term colonization is used for bacterial community which is multiflying within the wound but not causing tissue damage. The "critical colonization" is used to describe bacteria that are growing inside the wound but do not possess classical symptoms of infection. However, they adversely affect wound healing. Some researchers have suggested that bacteria may play a vital role in normal wound healing. However, the specific role of bacterial community in wound healing is still under debate. The wound communities have polymicrobial nature. Bacteria can find its way to a wound from exogenous (soil and water) and endogenous (skin, saliva, urine, and faeces) sources. Bacteria that do multiply are not considered "infective" unless and until they pose detrimental effects. This is particularly the case for Corynebacteria and coagulase-negative Staphylococcus which are considered as commensal skin organisms.

Biofilm Formation in Acute Wound

Biofilm formation in wounds has been investigated in-vivo using model organisms (murine and porcine). Akiyama and co-workers investigated S. aureus and Streptococcus biofilm formation in mice wounds. These mice were treated with cyclophosphamide to inhibit leukocytes. Normal mice rapidly clear the inoculated bacteria because of a strong PMN response. P. aeruginosa biofilm, after examining burn wounds of murine model, shows that P. aeruginosa rapidly colonized burn wounds and formed biofilms around blood vessels .

Biofilm Formation in Chronic Wound

Bacterial biofilm interrupts the human immune system in several ways. In initial stages of chronic wound, antibiotic treatment is considered to be the immediate step. But in case of mature and established biofilm, antibiotic therapy is least effective and has only short term effects on both inflammation and healing. Clinicians have to depend on the results from a swab or biopsy, which barely represent all microorganisms present in the wound. The bacterial community residing in biofilm can be up to 1000 times resistant to antimicrobials. Even silver treatment which is considered to be quite effective now a days and incorporated in wound dressings, is least effective on biofilm.

Biofilm in Infectious Kidney Stones

About 15-20% kidney stones are responsible for urinary tract infection and these stones are formed as a result of interaction between bacteria and mineral substances derived from urine. This result in the formation of complex biofilm composed of infecting bacteria and their exoproducts, and mineralized stone material. Hellstrom in 1938 for the first time examined the stones that were collected from his patients and discovered the occurrence of bacteria deep inside them. Microscopic analysis of stones, which were removed from infected patients, have revealed that bacteria present in the stones are organized to form microcolonies surrounded by an anionic matrix composed of complex polysaccharides and minerals. Urine flow is obstructed by these infectious stones and thus causes severe inflammation and infection that can lead to kidney failure.

Bacterial Endocarditis

Bacteria and host components form complex biofilm that cause infection lesion in endocarditis. This biofilm is known as vegetation and can cause disease by three main mechanisms . First, the vegetation disrupts the function of valve by creating leakage, turbulence and flow of blood. Secondly, the vegetation causes bloodstream infections, and may lead to recurrent fever, chronic systemic inflammation, and other severe complications. Thirdly, sometimes vegetation breaks off into pieces and these pieces are then carried to extremities in the circulation system (embolization). Brain and kidneys are particularly vulnerable areas of the body. Vegetation is usually treated by prolonged administration of intravenous antibiotics or surgical excision of the infected valve.

Cystic Fibrosis Airway Infections

Cystic Fibrosis patients most commonly afflicted with P. aeruginosa airway infections. These infections are generally divided into two steps. First, intermittent respiratory infections are developed in CF patients. In second stage, permanent infections with P. aeruginosa take place which last for the rest of the patient's life, as confirmed by genetic fingerprinting. This persistant infection is clinically very important as it causes permanent failure of lungs.

Other Biofilm Diseases

Biofilms cause several diverse kinds of human infections. Biofilm is involved in otitis media with effusion. Fluid accumulates in the middle ear cavity of the patient, thus affecting speech development and learning capability of the patient. However, the complete etiology of the problem is still not clealy understood. In acute osteomyelitis, certain areas of bone necrose and produce favourable conditions for biofilm development. Biofilms have also been identified in most indwelling medical device infections and also in biliary tract infections, periodontitis, ophthalmic infections.

Immune Response to Biofilms

In-vitro studies have indicated that human leukocytes have the ability to penetrate biofilms of S. aureus. Akiyama reported that antimicrobial efficacy against S. aureus biofilms was comparatively more effective in studying acute wound infection in normal mice than those who had less leucocytes count. They investigated the primary mechanism of antimicrobials in normal mice, and found that it was due to the penetration of PMNs into the biofilm. In P. aeruginosa biofilms, production of the extracellular polysaccharide, alginate, protected them from IFN-γ mediated phagocytosis by human leukocytes, primarily known as monocytes. Similarly, polysaccharide intercellular adhesin (PIA) protected S. epidermidis against killing by polymorphonuclear leukocytes (PMN) and phagocytosis. Thus, extracellular polysaccharides play a significant role in biofilm resistance to phagocytosis. In addition, P. aeruginosa biofilms cause killing of PMN through the production of rhamnolipids. Thus limits the effectiveness of innate immune factors. Severe cutaneous infections are caused by S. aureus which is capable of producing leukotoxins, including the Panton-Valentine leukocidin.

Pilicides

Pili are extracellular fibers of bacterial cells which allow binding and colonization on epithelial cells. These pili are assembled via a specific mechanism called chaperone-usher pathway. Interupting this mechanism of pili assembly is a new and feasible approach due to its potential application in biofilm eradication. Researchers have also designed small synthetic compounds known as pilicides which inhibit the synthesis of pili.

Enzymes

Another effective way to degrade biofilm is the application of enzymes. Biofilm consists of extracellular polymeric substances (EPS, therefore these enzymes have the potential to degrade EPS. Biofilm is mainly composed of bacteria and EPS. When the biofilm is degraded by enzymes which results release of components and plaktonic cells which is easily clear by immune systems.

Inhibition of Quorum-sensing

Using inhibitors of quorum sensing is the most studied novel way in control of biofilm. In last decade, many researchers have searched for compounds that could block the QS system.

Electrical Currents

It has been shown that electrical current has antibacterial effect for several bacterial species. In a study by the application of low intensity current, a substantial reduction was observed in number of both viable bacteria of staphylococcus and pseudomonas biofilms. Electrical currents in combination with electromagnetic fields and ultrasound have given enhanced results on biofilm eradication in studies conducted in-vitro as well as in-vivo.

Surface Coatings

One of the most effective ways for eradication or blocking bacterial biofilm on surfaces of endotracheal tubes (ETTs) and catheters is coating the devices with metals, antiseptics or antimicrobials.

Bacteriophages

Use of bacteriophages for eradication/removal of biofilm is an efficient and novel strategy . Bacteriophages ability to inhibit or reduce formation of biofilm in-vivo has also been proven . During a study, when a genetically engineered lytic phage having a biofilm degrading enzyme, was used showed more efficient eradication of biofilm than wild type phages. Similarly a phage cocktail (combination of multiple phages) can also be used for efficient and complete eradication of bacterial biofilm.

Bacterial Conjugation

Bacterial conjugation is a way by which a bacterial cell transfers genetic material to another bacterial cell. The genetic material that is transferred through bacterial conjugation is a small plasmid, known as F-plasmid (F for fertility factor), that carries genetic information different from that which is already present in the chromosomes of the bacterial cell. In fact, the F-plasmid can replicate in the cytoplasm separately from the bacterial chromosome.

A cell that already has a copy of the F-plasmid is called an F-positive, F-plus or F$^+$ cell, and is considered a donor cell, while a cell that does not have a copy of the F-plasmid is called an F-negative, F-minus or F$^-$ cell, and is considered a recipient cell. The transfer of the F-plasmid takes place through a horizontal connection by which the donor cell and the recipient cell directly contact each other or form a bridge between the two through which the genetic material is transferred. In cases where the F-plasmid of a donor cell has been integrated in the cell's genome (i.e., in the chromosome), a part of the chromosomal DNA may also be transferred to the recipient cell together with the F-plasmid.

Bacterial Conjugation Steps

In order to transfer the F-plasmid, a donor cell and a recipient cell must first establish contact. At this point, when the cells establish contact, the F-plasmid in the donor cell is a double-stranded DNA molecule that forms a circular structure. The following steps allow the transfer of the F-plasmid from one bacterial cell to another:

Step 1

The F$^+$ (donor) cell produces the pilus, which is a structure that projects out of the cell and begins contact with an F$^-$ (recipient) cell.

Step 2

The pilus enables direct contact between the donor and the recipient cells.

Step 3

Because the F-plasmid consists of a double-stranded DNA molecule forming a circular structure, i.e., it is attached on both ends, an enzyme (relaxase, or relaxosome when it forms a complex with other proteins) nicks one of the two DNA strands of the F-plasmid and this strand (also called T-strand) is transferred to the recipient cell.

Step 4

In the last step, the donor cell and the recipient cell, both containing single-stranded DNA, replicate this DNA and thus end up forming a double-stranded F-plasmid identical to the original F-plasmid. Given that the F-plasmid contains information to synthesize pili and other proteins, the old recipient cell is now a donor cell with the F-plasmid and the ability to form pili, just as the original donor cell was. Now both cells are donors or F+.

The four steps mentioned above can be seen in this figure:

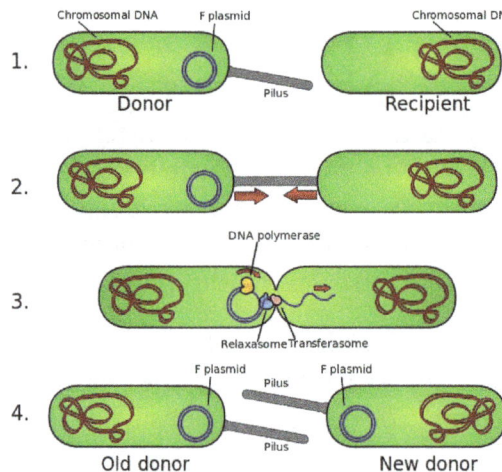

DNA Transfer

In order to avoid transferring the F-plasmid to an F+ cell, the F-plasmid usually contains information that allows the donor cell to detect (and avoid) cells that already have one. In addition, the F-plasmid contains two main loci (tra and trb), an origin of replication (OriV) and an origin of transfer (OriT). The tra locus contains the genetic information to enable the donor cell to be attached to a recipient cell: the genes in the tra locus code for proteins to form the pili (pilin gene) in order to start the cell-cell contact, and other proteins to get attached to the F− cell and to start the transfer of the F-plasmid. The trb locus contains DNA that codes for other proteins, such as some that are involved in creating a channel through which the DNA is transferred from the F+ to the F− cell. The OriV is the site at which replication of the DNA occurs and the OriT is the site at which the enzyme relaxase (or the relaxosome protein complex) nicks the DNA strand of the F-plasmid.

Although the DNA that is transferred in bacterial conjugation is that present in the F-plasmid, when the donor cell has integrated the F-plasmid into its own chromosomal DNA, bacterial conjugation can result in the transfer of the F-plasmid and of chromosomal DNA. When this is the case, a longer contact between the donor and the recipient cells results in a larger amount of chromosomal DNA being transferred.

The advantages of bacterial conjugation make this method of gene transfer a widely used technique in bioengineering. Some of the advantages include the ability to transfer relatively large sequences of DNA and not harming the host's cellular envelope. Furthermore, conjugation has been achieved in laboratories not only between bacteria, but also between bacteria and types of cells such as plant cells, mammalian cells and yeast.

References

- Bacterial-Genetics: biologyreference.com, Retrieved 16 March 2018

- Bacterial-growth: golifescience.com, Retrieved 28 May 2018

- How-microbes-grow, microbiology: courses.lumenlearning.com, Retrieved 11 July 2018

- Bacterial-conjugation: biologydictionary.net, Retrieved 14 May 2018

- Bacterial-biofilm-its-composition-formation-and-role-in-human-infections-61426: rroij.com, Retrieved 20 June 2018

Chapter 5

Bacterial Classification

Bacteria share several common characteristics, such as unicellularity, lack of nuclear membrane, etc. However, there exist differences in terms of morphology, metabolism, phylogeny, pathogenicity, etc. thus allowing their identification and classification. This chapter covers all the significant classifications of bacteria and also discusses the classification on the basis of flagella, spore, gaseous requirement, etc.

Classification on the Basis of Nutrition

The four major nutritional are:

Photoautotrophs

These bacteria capture the energy of sun light and transform it into the chemical energy. In this process CO_2 is reduced to carbohydrates. The hydrogen donor is water and the process produce free oxygen. Chlorophyll is present in the cell and its main function is to capture sun light e.g., Cyanobacteria.

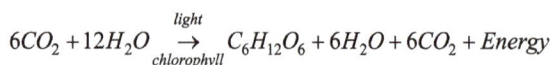

$$6CO_2 + 12H_2O \xrightarrow[chlorophyll]{light} C_6H_{12}O_6 + 6H_2O + 6CO_2 + Energy$$

The reaction produces free oxygen. However, in some bacterial photosynthesis hydrogen donor is a substance other than water, hence, oxygen is never produced. This is called an oxygenic photosynthesis and is found in purple sulphur bacteria and green sulphur bacteria.

This photoautotrophic bacteria are anaerobes and have bacteriochlorophyll and bacteriovirdin pigments respectively. In the process of photosynthesis these pigments absorb light and reduce carbon dioxide to form organic compounds. Carbon dioxide is taken from the atmosphere and hydrogen from sources except water.

Purple Sulphur Bacteria

These bacteria have the pigment bacteriochlorophyll located on the intracytoplasmic membrane i.e., thylakoids. These bacteria obtain energy from sulphur compounds e.g., Chromatiiun. Theopedia rosea, Thiospirilium.

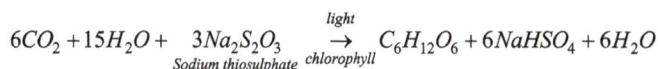

$$6CO_2 + 15H_2O + \underset{Sodium\ thiosulphate}{3Na_2S_2O_3} \xrightarrow[chlorophyll]{light} C_6H_{12}O_6 + 6NaHSO_4 + 6H_2O$$

Green Sulphur Bacteria

These bacteria used hydrogen sulphide (H_2S) as hydrogen donor. The reaction takes place in the presence of light and pigment termed as bacteriovirdin or bacteriopheophytin or chlorobium chlorophyll e.g., Chlorobium limicola, Chlorobacterium etc.

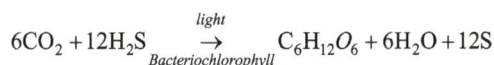

$$6CO_2 + 12H_2S \xrightarrow[Bacteriochlorophyll]{light} C_6H_{12}O_6 + 6H_2O + 12S$$

These bacteria take hydrogen from inorganic sources like sulphides and thiosulphates. Therefore, these bacteria are also known as photolithographs.

Photoheterotrophs (Gr., Photo = light; hetero = (an), other; troph = nourishment)

These bacteria can trap light energy but cannot use carbon dioxide as their sole caron source. They use organic compounds from the environment to satisfy their carbon and electron requirements. They use organic compounds such as carbohydrates, fatty acids and alcohols as their organic food. The pigment is bacteriochlorophyll e.g., Purple non-sulphur bacteria (Rhodospirillum, Rhodomicrobium, Rhodopseudomonas palustris).

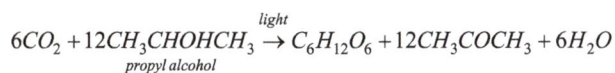

$$6CO_2 + 12CH_3CHOHCH_3 \xrightarrow{light} C_6H_{12}O_6 + 12CH_3COCH_3 + 6H_2O$$
$$\text{propyl alcohol}$$

Chemoautotrophs

These bacteria do not require light (lack the light phase but have the dark phase of photosynthesis) and pigment for their nutrition. These bacteria oxidize certain inorganic substances with the help of atmospheric oxygen. This reaction releases the energy (exothermic) which is used to drive the synthetic processes of the cell.

The source of carbon is carbon dioxide. In the absence of light synthesis of organic food from inorganic substances by utilizing chemical energy is also known as chemosynthesis. The chemoautotrophic bacteria play a very important role in recycling inorganic nutrients. These bacteria are commonly named after the structure of the compound which is utilized as the source of energy viz.

Nitrifying Bacteria

These bacteria obtain energy by oxidizing ammonia into nitrate. The process occurs in two steps and each step is carried out by a specialized group of bacteria.

In the first step ammonia is oxidized into nitrites by the bacteria Nitrosomonas, Nitrococcus:

$$2NH_3 + 3O_2 \rightarrow 2HNO_2 + 2H_2O + 79 \; k.cal$$
$$\text{Nitrate}$$

In the second step, the nitrites are converted into nitrate.

This is brought about by the bacteria Nitrobacter, Nitrocystis.

$$2HNO_2 + O_2 \rightarrow 2HNO_3 + 21.6 \; k.cal$$

Nitrite Nitrate

These nitrifying bacteria are present in the soil and are of considerable economic importance. When these two groups of bacteria work together, ammonia in the soil is oxidized to nitrate in a process called nitrification. Energy released upon the oxidation of both ammonia and nitrite is used for chemosynthesis by these bacteria (to make ATP by oxidative phosphorylation).

Sulphur Bacteria

These bacteria obtains energy either by oxidation of elemental sulphur or H_2S.

Elemental Sulphur Oxidising Bacteria

Denitrifying sulphur bacteria e.g., Thiobacillus denitrificans oxidize elemental sulphur to sulphuric acid and utilize energy produced in this process.

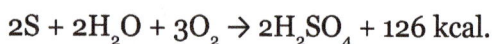

$$2S + 2H_2O + 3O_2 \rightarrow 2H_2SO_4 + 126 \text{ kcal.}$$

Sulphide Oxidizing Bacteria

These bacteria oxidizes H_2S and release the sulphur e.g., Beggiatoa.

$$2H_2S_4 - O_2 \rightarrow 2H_2O + 2S + 141.8 \text{ cal}$$

Sulphuric acid

Iron Bacteria

These bacteria inhabit waters that contain inorganic iron compounds and oxidize ferrous compounds to ferric forms e.g., Thiobacillus ferroxidans, Ferro bacillus, Leptothrix.

$$4FeCo_3 + 6H_2O + O_2 \rightarrow 4Fe(OH)_3 + 4CO_2 + 81 \text{ kcal.}$$

Hydrogen Bacteria

These bacteria oxidizes hydrogen into water e.g. Hydrogenomonas.

$$2H_2 + O_2 \rightarrow 2H_2O + 55 \text{ kcal.}$$

$$4H_2 + CO_2 \rightarrow 2H_2O + CH_4 + \text{Energy}$$

Carbon Bacteria

These bacteria oxidizes CO into CO_2 e.g., Bacillus oligocarbophillous, Oligotropha carboxydovorans

$$2CO + O_2 \rightarrow 2CO_2 + \text{Energy}$$

Chemoheterotrophs

These bacteria obtain both carbon and energy from organic compounds such as carbohydrates, lipids and proteins. The carbon source as well as the source of energy is mostly the same for these bacteria. Most of the bacteria are chemo heterotrophs.

Glucose or Monosaccharide $[(CH_2O)_n] + O_2 \rightarrow CO_2 + H_2O +$ Energy.

Chemo heterotrophs may belong to one of the three main categories that differ in how they obtain their organic nutrients.

These are

(i) Parasitic

These bacteria obtain their food from living hosts on which these grow. Parasites which cause diseases are known as pathogens e.g., Clostridium, Mycobacterium etc.

(ii) Saprophytic

These bacteria obtain their food from dead and organic remains like fruits, vegetables, leaves,meat, faeces, corpses and other non-living products. The anaerobic breakdown of carbohydrates is fermentation while that of proteins is called putrefaction, e.g., Putrefying bacteria like Bacillus mycoides, B. ramosus etc.

(iii) Symbiotic

These bacteria live in close association with organs of other organisms (higher plants and animals) in such a way that both the concerned organism receive mutual benefit from this association. This is called symbiosis for e.g., Rhizobium leguminosarum in the root nodules of the leguminous plants.

This bacteria fix free atmospheric nitrogen into nitrogenous compounds which are utilized by the plants. In return, the plant provides nutrients and protection to the bacteria. In the stomach of the cows and goats bacteria digest cellulose enabling these animals to feed on grass. Our own intestine contains a number of harmless bacteria e.g., Escherichia coli.

Classification on the Basis of Temperature

Psychrophiles

- Bacteria that can grow at 0°C or below but the optimum temperature of growth is 15°C or below and maximum temperature is 20°C are called psychrophiles.
- Psychrophiles have polyunsaturated fattyacids in their cell membrane which gives fluid nature to the cell membrane even at lower temperature.
- Examples: Vibrio psychroerythrus, vibrio marinus, Polaromonas vaculata, Psychroflexus.

Psychrotrops (Facultative Psychrophiles)

- Those bacteria that can grow even at 0°C but optimum temperature for growth is (20-30) °C.

Mesophiles

- Those bacteria that can grow best between (25-40)°C but optimum temperature for growth is 37°C.

- Most of the human pathogens are mesophilic in nature.

- Examples: coli, Salmonella, Klebsiella, Staphulococci.

Thermophiles

- Those bacteria that can best grow above 45°C.

- Thermophiles capable of growing in mesophilic range are called facultative thermophiles.

- True thermophiles are called as Stenothermophiles, they are obligate thermophiles.

- Thermophils contains saturated fattyacids in their cell membrane so their cell membrane does not become too fluid even at higher temperature.

- Examples: Streptococcus thermophiles, Bacillus stearothermophilus, Thermus aquaticus.

Hypethermophiles

- Those bacteria that have optimum temperature of growth above 80°C.

- Mostly Archeobacteria are hyperthermophiles.

- Monolayer cell membrane of Archeobacteria is more resistant to heat and they adopt to grow in higher remperature.

- Examples: Thermodesulfobacterium, Aquifex, Pyrolobus fumari, Thermotoga.

Classification on the Basis of pH Growth

Different microorganisms often require distinct environments, with varied temperature, levels of oxygen, light and acidity or pH level. Some microbes grow faster in environments with extremely low pH values. These are called acidophiles, because of their preference for acidic environments. Although most microorganisms require neutral pH values to have optimum growth, alkaliphilic microorganisms prefer low-acidity or high pH environment.

Acidophiles

Microorganisms which optimum growth at pH levels lowers than 5 are called acidophiles. These microbes are found in a variety of environments, including geysers and sulfuric pools, as well as in the human stomach. Examples of acidophiles include the microscopic algae Cyanidium caldarium and Dunaliella acidophila. The microscopic fungi, Acontium cylatium, Cephalosporium and Trichosporon cerebriae can grow near pH 0. A primitive microorganism called Picrophilaceae have optimum pH values close to zero and can also grow at negative pH values.

Alkaliphilic

Alkaliphilic microorganisms have optimum growth at pH values between 9 and 12. These microorganisms thrive in alkaline lakes, soils and other high pH environments. In slag dumps of Lake Calumet, southeast Chicago, the water can reach a pH of 12.8, which is similar to caustic soda. Some bacteria related to the Clostridium and Bacillus live in that extremely alkaline environment. Mono Lake, in California, and Octopus Spring in Yellowstone Park are examples of environments were alkaliphilic microorganisms are found.

Neutrophiles

Neutral pH values, laying between 6 and 8, are more commonly found in nature. Along their evolution, most microorganisms have adapted to have optimum growths in acidity-neutral environments. These microorganisms are called neutrophiles, and include most species of microalgae and other organisms that form the phytoplankton, as well as some soil-dwelling bacteria and yeasts.

Classification on the Basis of Salt Requirement

Halophile

Halophilic extremophiles, or simply halophiles, are a group of microorganisms that can grow and often thrive in areas of high salt (NaCl) concentration. These hypersaline areas can range from the salinity equivalent to that of the ocean (~3-5%), up to ten times that, such as in the Dead Sea (31.5% average). Halophiles have been found belonging to each domain of life but primarily consist of archaea. They are metabolically diverse, ranging from simple fermenters to iron reducers and sulfide oxidizers.

Physical Environment

Naturally-Occurring Hypersaline Environments

Figure: Salterns changing color with increasing salinity, due to various pigments different halophiles produce. These pigments facilitate UV resistance and/or phototrophy.

Hypersaline environments are present on each continent and are primarily found in "arid and semi-arid regions." Some are commonly known, like Utah's Great Salt Lake, or the Dead Sea between Israel and Jordan. Others are less well known, such as Antarctica's Deep Lake or Papua New Guinea's undersea geothermal vents.

Constructed Hypersaline Environments

Anthropogenic hypersaline environments are commonly created by the salt industry. Salterns are large ponds that are filled with saltwater from the ocean or another source that are then evaporated away. In this process, the salinity of the water gradually increases as water evaporates until it reaches saturation (26% at 20°C). The salt then precipitates out and is harvested. Halophiles take advantage of this environment and often their presence becomes visible due to pigments they produce.

Conditions

Salt

	Halophile Classification		
	Slight Halophile	Moderate Halophile	Extreme Halophile
Percent Salt	2 - 5%	5 - 20%	20 - 30%
Molarity	0.34 - 0.85 M	0.85 - 3.4 M	3.4 - 5.1 M

Often in hypersaline environments, salts other than NaCl are also present. A salt profile for the Dead Sea is shown below. However, because halophiles are defined in relation to NaCl concentration, other salt content is not considered for halophilic classification.

Salt Profile of the Dead Sea							
Salt Type	$MgCl_2$	NaCl	$CaCl_2$	KCl	$MgBr_2$	$CaSO_4$	RbCl
Percent Total Salts	51.26%	28.19%	13.64%	4.57%	2.17%	0.10%	0.03%

	Salinity and Temperature Measurements for Saline Environments.			
	Pacific Ocean	Deep Lake, Antarctica	Great Salt Lake, USA	Dead Sea, Israel/Jordan
NaCl Percent (g/L)	3.4 - 3.7%	21 - 28%	12 - 33%	31.5 - 34.2%
Temperature (C)	1.4 - 30 C	0 - 11.5 C	-5 - 35 C	21 - 36 C

Temperature

Often in hypersaline environments, the salinity is just one extreme microbes must overcome. Hypersalinity often co-occurs with extreme temperature conditions, both hot and cold.

Psychrophiles

In the case of Deep Lake, Antarctica, extreme cold and high salinity meet. Deep Lake has salinity levels ranging from 21-28%, putting the halophiles present in the extreme halophile classification. Psychrophiles are "organisms having an optimal temperature for growth at about 15°C or lower, a maximal temperature for growth at about 20°C, and a minimal temperature for growth at 0°C or below".[11] In the case of Deep Lake, eight months of the year are spent below 0°C, with a yearly maximum temperature of only 11.5°C.[6] Organisms that survive and thrive here are not only halophiles, but also psychrophiles. Similar overlaps of cold and hypersaline environments have been found in other places as well.

Figure: An image of Deep Lake, Antarctica.[25]

Thermophiles

In the case of the black smokers off the coast of Papua New Guinea, a hypersaline environment was created by a hydrothermal vent. This hydrothermal vent releases a hypersaline effluent that was measured at greater than 250°C. Samples taken from the internal walls of the smoker contained *Halomona* and *Haloarcula* species.[19] This classifies *Halomona* and *Haloarcula* as both thermophiles and halophiles. Such overlapping conditions of hypersalinity and extreme heat are also present in some hot springs and other hydrothermal vents.

Other Extremes

Some hypersaline environments have been found that overlap with other extremes, such as low and high pH, and dry, desiccating conditions. Organisms in such conditions would be considered haloacidophiles, haloalkaliphiles, and haloxerophiles, respectively.

Halotolerance is the adaptation of living organisms to conditions of high salinity. Halotolerant species tend to live in areas such as hypersaline lakes, coastal dunes, saline deserts, salt marshes, and inland salt seas and springs. Halophiles are organisms that live in highly saline environments, and require the salinity to survive, while halotolerant organisms (belonging to different domains of life) can grow under saline conditions, but do not require elevated concentrations of salt for growth. Halophytes are salt-tolerant higher plants. Halotolerant microorganisms are of considerable biotechnological interest.

The extent of halotolerance varies widely amongst different species of bacteria. A number of cyanobacteria are halotolerant; an example location of occurrence for such cyanobacteria is in the Makgadikgadi Pans, a large hypersaline lake in Botswana.

Classification on the Basis of Gaseous Requirement

Aerobes: Aerobe grow in ambient air, which contains 21% oxygen and small amount of (0.03%) of carbon dioxide. Aerobes require molecular oxygen as a terminal electron acceptor so cannot grow in its absence. e.g., Bacillus cereus.

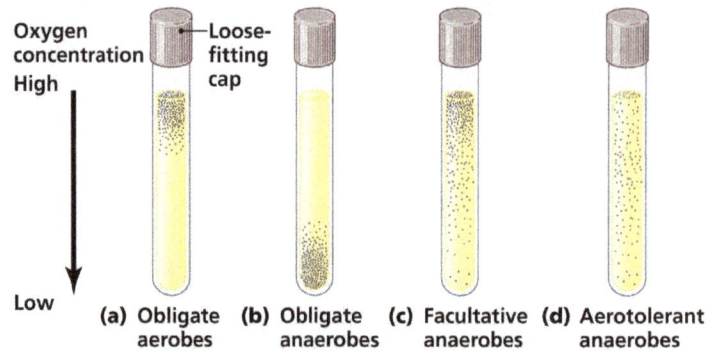

(a) Obligate aerobes (b) Obligate anaerobes (c) Facultative anaerobes (d) Aerotolerant anaerobes

Obligate aerobes: They have absolute requirement for oxygen in order to grow. Pseudomonas aeruginosa, Mycobacterium tuberculosis.

Anaerobes: Usually bacteria of this group cannot grow in the presence of oxygen, oxygen is toxic for them. They use other substances as terminal electron acceptor. Their metabolism frequently is a fermentative type in which they reduce available organic compounds to various end products such as organic acids and alcohols.

Obligate anaerobes: These bacteria grow only under condition of high reducing intensity and for which oxygen is toxic. Clostridium perfringens, Clostridium botulinum etc.

Facultative anaerobes: They are versatile organisms, capable of growth under both aerobic and anaerobic conditions. They preferentially use oxygen as terminal electron acceptor. e.g., Enterobacteriaceae group, Staphylococcus aureus etc.

Aerotolerant anaerobes: Are anaerobic bacteria that are not killed by exposure to oxygen.

Capnophiles: Capnophilic bacteria require increased concentration of carbon dioxide (5% to 10%) and approximately 15% oxygen. This condition can be achieved by a candle jar (3% carbon dioxide) or carbon dioxide incubator, jar or bags. The examples of capnophilic bacteria includes: Haemophilus influenzae, Neisseria gonorrhoeae etc.

Microaerophiles: Microaerophiles are those groups of bacteria that can grow under reduced oxygen (5% to 10%) and increased carbon dioxide (8% to 10%). Higher oxygen tensions may be inhibitory to them. This environment can be obtained in specially designed jars or bags. Examples of Microaerophiles are: Campylobacter jejuni, Helicobacter pylori etc.

GRAM POSITIVE BACTERIA

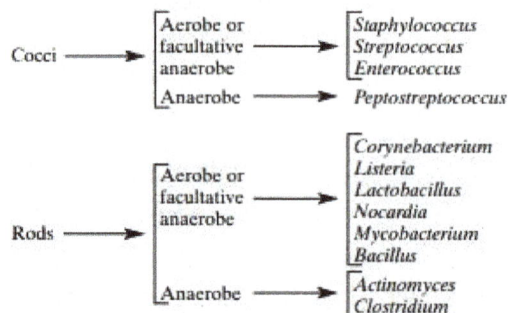

GRAM NEGATIVE BACTERIA

```
Cocci ──────► Aerobe ──────────► Neisseria

              ┌ Aerobe ──────────► Pseudomonas
              │                    ┌ Salmonella
              │                    │ Shigella
              │                    │ Klebsiella
              │                    │ Proteus
              │ Facultative        │ Escherichia
Rods ───────► │ anaerobe ────────► │ Yersinia
              │                    │ Bordetella
              │                    │ Haemophilus
              │                    │ Brucella
              │                    │ Pasteurella
              │                    └ Vibrio
              │                    ┌ Bacteroides
              │ Anaerobe ────────► │ Fusobacterium
              │                    └ Prevotella
              └ Microaerophile ──► Campylobacter

              ┌ Aerobe ──────────► Leptospira
Spirochaetes ►│                    ┌ Borrelia
              └ Anaerobe ────────► └ Treponema
```

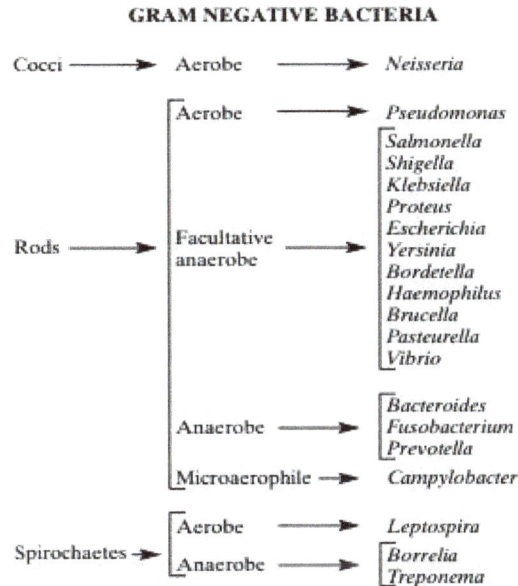

Basic classification of Medically Important Bacteria

Classification	Characteristics	Important Genera
Obligate aerobes	Require oxygen, Have no fermentative pathways. Generally produce superoxide dismutase	Mycobacterium Pseudomonas Bacillus
Microaerophilic	Requires low but not full oxygen tension	Campylobacter Helicobacter
Facultative anaerobes	Will respire aerobically until oxygen is depleted and then ferment or respire anaerobically	Most bacteria,i.e., Enterobacteria-ceae
Obligate anaerobes	Lack superoxide dismutase Generally lack catalase Are fermenters Can not use oxygen as terminal electron acceptor	Actinomyces* Bacteroides Clostridium *Mneomonics: ABCs of anaerobiosis

Classification on the Basis of Morphology

1. Coccus

- These bacteria are spherical or oval in shape

- On the basis of arrangement, cocci are further classified as:

  o Diplococcus: coccus in pair. Eg, *Neissseria gonorrhoae, Pneumococcus*

  o Streptococcus: coccus in chain. Eg. *Streptococcus salivarius*

  o Staphylococcus: coccus in bunch. Eg. Staphylococcus aureus

  o Tetrad: coccus in group of four

  o Sarcina: cocus in cubical arrangement of cell. Eg. *Sporosarcina*

2. Bacilli

- These are rod shaped bacteria
- On the basis of arrangement, bacilli are further classified as:
    - o Coccobacilli: Eg. *Brucella*
    - o Streptobacilli: chain of rod shape bacteria: Eg. *Bacillus subtilis,*
    - o Comma shaped: Eg. *Vibrio cholarae*
    - o Chinese letter shaped: *Corynebacterium dephtherae*

3. Mycoplasma

- They are cell wall lacking bacteria
- Also known as PPLO (Pleuropneumonia like organism)
- *Mycoplasma pneumoniae*

4. Spirochaetes

- They are spiral shaped bacteria
- Spirochaetes

5. Rickettsiae and Chlamydiae

- They are obligate intracellular parasites resemble more closely to viruses than bacteria

6. Actinomycetes

- They have filamentous or branching structure
- They resemble more closely to Fungi than bacteria
- Example: *Streptomyces*

Classification on the Basis of Gran Staining

Gram Positive Bacteria

Gram positive bacteria have a thick multilayered, peptidoglycan cell wall that is exterior to the membrane. The peptidoglycan in most gram positive bacteria is covalently linked to teichoic acid, which is essentially a polymer of substituted glycerol units linked by phosphodiester bonds. All gram positive bacteria also have teichoic acid in their membranes, where it is covalently linked to glycolipid. The teichoic acids are major cells surface antigens. The cell wall structure and example of gram positive will be shown in figures below.

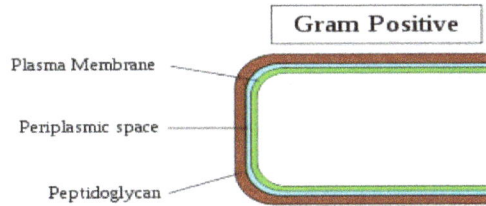

Figure: Gram positive cell wall structure.

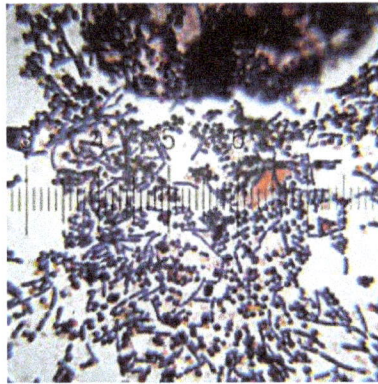

Figure: Gram-positive bacteria, stained purple, of both the bacillus ("rodshaped") and cocci (spherical) forms. A few Gram-negative bacteria are also present, stained pink.

Gram Negative Bacteria

Gram negative bacteria have two membranes – an outer membrane and inner (cytoplasmic) membrane. Their peptidoglycan layer is located between the two membranes in what is called the periplasmic space. The periplasmic space also contains enzymes and various other substances. In contrast to gram positive cells, the peptidoglycan layer of gram negative is thin, and the cells are consequently more susceptible to physical damage. The outer membrane is distinguished by the presence of various embedded lipopolysaccharides. The polysaccharide portion (Opolysaccharide) is antigenic, and can therefore be used to identity different strains and species. The lipid portion (lipid A) is toxic to humans and animals. Lipid A, because it is an integral part of membrane, is called an endotoxin, as opposed to exotoxins, which are secreted substances. The cell wall structure and the example of gram negative will be shown in figures below.

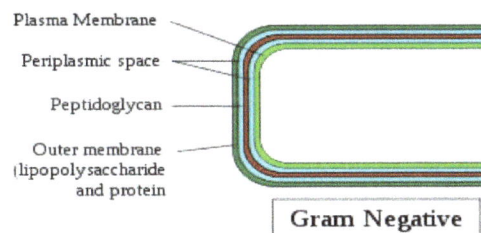

Figure: Gram negative cell wall structure.

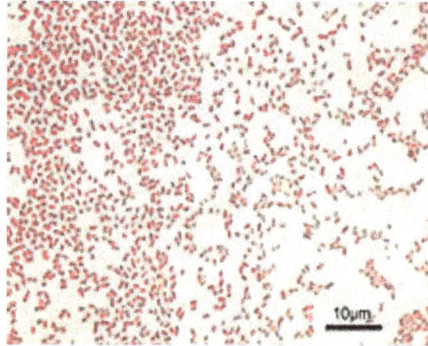

Figure: Microscopic image of Gram-negative Pseudomonas aeruginosa bacteria (pink-red rods).

Figure: shows the sputum sample image taken under x10 magnification using digital microscope that used for detection and summation of pus cell and squamous epithelial cells.

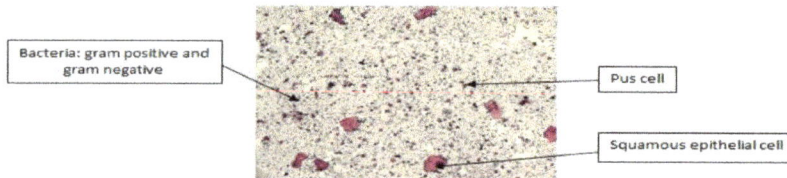

Figure: Sputum image under x10 computerized microscope.

Figure: shows the sputum sample image taken under x100 magnification using digital microscope that used for detection and summation of bacteria (gram positive and gram negative).

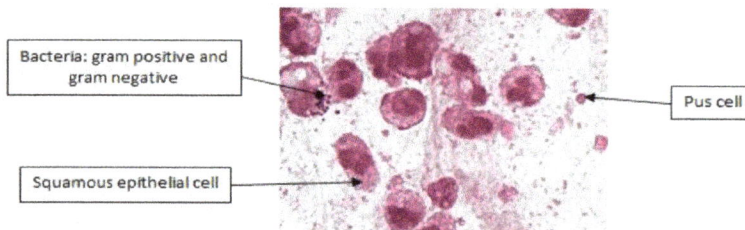

Figure: Sputum image under x100 computerized microscope.

Classification on the Basis of Flagella

On the basis of flagella the bacteria can be classified:

Monotrichous

- Single polar flagellum
- Example: Vibrio cholerae

Amphitrichous

- Single flagellum on both sides
- Example: Alkaligens faecalis

Lophotrichous

- Tufts of flagella at one or both sides
- Example: Spirillum

Peritrichous

- Numerous falgella all over the bacterial body
- Example: Salmonella Typhi

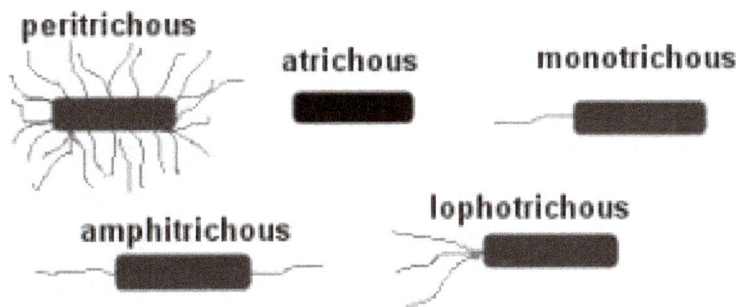

Classification on the Basis of Spore

Spore Forming Bacteria

- Those bacteria that produce spore during unfavorable condition.
- These are further divided into two group

Endospore Forming Bacteria

- Spore produced within the bacterial cell
- Bacillus, Clostridium, Sporosarcina etc.

Bacillus: Anthrax and Research

Bacillus is a genus of spore forming, aerobic, rod-shaped bacteria. This fairly large group is most notorious for Bacillus anthracis, the bacterium responsible for the deadly disease anthrax. According to the Centers for Disease Control and Prevention, the bacterium's spores allow it to last a long time in the environment before entering people and causing an infection. But another member of this genus, Bacillus subtilis, is commonly used by molecular biology researchers to investigate basic questions of gene regulation and the cell cycle. Other Bacillus species include Bacillus cereus, Bacillus clausii and Bacillus halodenitrificans. Some of these bacteria are common causes of food and medical contamination and can be difficult to eliminate.

Clostridium: Disease and Production

Clostridium forms spores that differ from those of other bacteria in being pin- or bottle-shaped, not the usual ovals. According to Public Health England, the Clostridium genus contains over 100 species, including harmful pathogens such as Clostridium botulinum, Clostridium difficile, Clostridium perfringens, Clostridium tetani and Clostridium sordellii. However, some species of the bacteria are used commercially to produce ethanol (Clostridium thermocellum) and acetone (Clostridium acetobutylicum), as well as to convert fatty acids to yeasts and propanediol (Clostridium diolis).

Sporolactobacillus: Lactic Acid Makers

Sporolactobacillus are unique among spore forming bacteria in also being lactic acid bacteria. These species, such as Sporolactobacillus dextrus, Sporolactobacillus inulinus, Sporolactobacillus laevis, Sporolactobacillus terrae and Sporolactobacillus vineae, make lactic acid as the end product of their metabolism. They mainly consume carbohydrates such as fructose, sucrose, raffinose, mannose, inulin and sorbitol.

Sporosarcina: Breaking Down Urine

Sporosarcina are a group of bacteria with both rod-shaped and round (coccoid) members. The most famous member of the genus, Sporosarcina ureae, is able to break down urea, the chemical that gives urine its distinctive smell. This bacterium is particularly common in soils that receive a lot of urine, such as fields underneath grazing cows. Other species in the genus include Sporosarcina aquimarina, Sporosarcina globispora, Sporosarcina halophila, Sporosarcina koreensis and Sporosarcina luteola.

Exospore Forming Bacteria

- Spore produced outside the cell
- Methylosinus

Non Sporing Bacteria

- those bacteria which do not produce spore
- Eg. E. coli, Salmonella

References

- Dieter Häussinger and Helmut Sies (2007) Osmosensing and Osmosignaling, Academic Press, 579 pages ISBN 0-12-373921-7

- Types-microorganisms-optimum-ph-8618232: sciencing.com, Retrieved 14 April 2018

- Oxygen-requirements-for-pathogenic-bacteria: microbeonline.com, Retrieved 28 March 2018

- Classification-of-bacteria: onlinebiologynotes.com, Retrieved 10 April 2018

- Flagella-introduction-types-examples-parts-functions-and-flagella-staining-principal-procedure-and-interpretation: microbiologyinfo.com, Retrieved 25 June 2018

- Types-spore-forming-bacteria-2504: sciencing.com, Retrieved 14 March 2018

Chapter 6

Bacterial Interactions

Bacteria form complex associations with other organisms in the form of symbiotic associations such as mutualism, parasitism and commensalism. This chapter explores the interactions of bacteria with the environment and with other organisms, through a discussion on host-pathogen interaction, transmission methods, bacterial infection methods, etc.

Host–pathogen Interaction

Host-pathogen interactions provide information that can help scientists and researchers understand disease pathogenesis, the biology of one or many pathogens, as well as the biology of the host. It is through these interactions that basic research discoveries are made.

Koch's Postulates: Experimental Steps to Determine Disease Causation

Over a century ago, Robert Koch established that infectious diseases were caused by microbes. He was looking for the causative agent for anthrax. Koch's postulates are experimental criteria that are used to determine if a microbe caused a specific disease. The criteria include:

1. The microbe must be present in every case of the disease.

2. The organism must be grown in a pure culture from diseased hosts.

3. The same disease must be produced when a pure culture of the organism is introduced into a susceptible host.

4. The organism must be recovered from the experimentally infected hosts.

However, there are some exceptions to these criteria. These include:

1. Some organisms cannot be cultured in a lab and grown on artificial media.

2. Some pathogens can cause several disease conditions such as M. tuberculosis, which can cause lung disease and other diseases of the skin, bone, and internal organs.

3. There may be ethical reasons that do not allow testing, (i.e., human diseases with no animal model – smallpox, rubella).

An individual's or animal's skin and mucous membranes generate an environment for microorganisms to interact with the body. This interaction between the host and the organism is referred to as symbiosis. There are three forms of symbiotic relationships that can occur at an anatomical level. In order for this to take place the following components will need to occur:

- Mutualism: In mutualism, both the microorganism and the body work together. An example of this relationship would be cows and the bacteria in their rumen. Bacterial cellulose facilitates digestion in the animal, while the bacteria benefit from nutrients in the rumen.

- Commensalism: In commensalism, either the body or the microorganism benefits, while the other is not affected by the interaction. Examples of this include microorganisms that make up the normal flora that inhabit the eyes. These organisms thrive on secretions and dead cells, but do not affect the host.

- Parasitism: In parasitism, one organism benefits at the expense of the other. For example, parasite use the gastrointestinal tract of a human or animal as an environment in which to reproduce.

The Body's Normal Flora

The body contains two types of normal flora: 1) resident flora (survive for extended periods), and 2) transient flora (temporary). Normal flora help to provide defenses against invading pathogens by covering adherence sites, producing compounds toxic to other organisms, and preventing pathogens from consuming available nutrients. For disease to occur there must be a change in the body's environment, which, in turn, allows the pathogen to overcome the normal flora. This can occur through a change in the pH of the body or elimination of normal flora due to antibiotics.

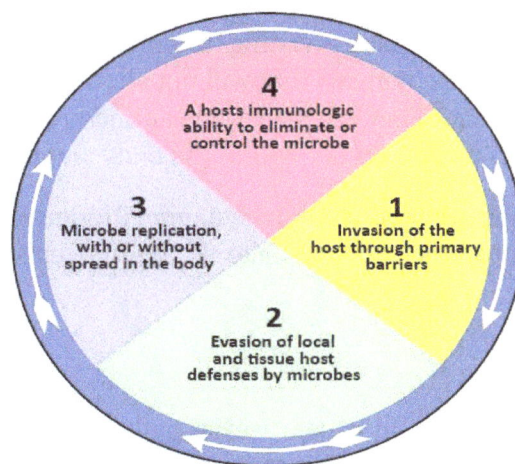

4 A hosts immunologic ability to eliminate or control the microbe

3 Microbe replication, with or without spread in the body

1 Invasion of the host through primary barriers

2 Evasion of local and tissue host defenses by microbes

Components for the host pathogen interaction.

Host Defenses

A microorganism will not be able to invade unless it overcomes an animal's or individual's host defenses. Specific host defenses may include:

- Skin and mucosal secretions

- Non-specific local responses (e.g., pH)

- Non-specific inflammatory responses

- Specific immune responses (e.g., lymphocytes)

The ability for a pathogen to overcome host defenses can be accomplished by two distinct components: a primary pathogen (causes disease in a healthy host) or opportunistic pathogen (causes disease if host is immunocompromised).

Pathogen Defenses

Pathogens contain virulence factors that promote disease formation and provide the opportunity for a microbe to infect and cause disease. The greater the virulence, the more likely disease will occur. Such factors include:

- Ability of a pathogen to adhere to a host

- Ability of a pathogen to colonize (overcome) a host

- Ability of a pathogen to evade host defenses

Mechanisms of Pathogenesis

Pathogenesis is the method by which a disease can develop. This can occur through foodborne intoxication where the causative agent produces toxins in the body (e.g., botulism). Another route is the colonization of an invading pathogen on the host surface, which allows the pathogen to increase in numbers and produce toxins that are damaging to the host's cells (e.g., Vibrio and Corynebacterium).

Pathogenesis can also occur by pathogens invading and breaching the body's barrier in order to multiply. These organisms have mechanisms that will not allow macrophages (the body's defense against pathogens) to destroy them. They can also evade antibody detection (e.g., tuberculosis and plague). Finally, organisms can invade tissues within the body and produce toxins (e.g., Shigella).

The relationship between a host and pathogen is dynamic. Production of disease occurs through a process of steps. The first five mechanisms make up a pathogen's invasiveness (i.e., ability to invade tissues).

Transmission

In order to begin infection and eventually cause disease, pathogens must find a transmission route. Transmission of an infectious agent can occur in many ways, but it is typically through exposed skin (e.g., a ways, but it is typically through exposed skin (e.g., a cut, abrasion, puncture, or wound) or mucous membranes (e.g., gastrointestinal tract, respiratory tract, or urogenital tract).

Adherence

Once the pathogen has gained access to the body, it must have some means of attaching itself to the host's tissues. This attachment is called adherence and is a necessary step in pathogenicity. Microbes contain ligands, which are projections that attach host receptors or surface proteins. Pathogens may have specific adherence mechanisms to attach to cells or tissue surfaces. Examples of this include: 1) tissue tropism (i.e., pathogens that prefer specific tissues over others), 2) species specificity (i.e., pathogens that only infect certain species), and 3) genetic specificity (i.e., surface mutations that occur so previous antibodies do not recognize the invading pathogen). If a microorganism cannot adhere to a host cell membrane, disease will not occur.

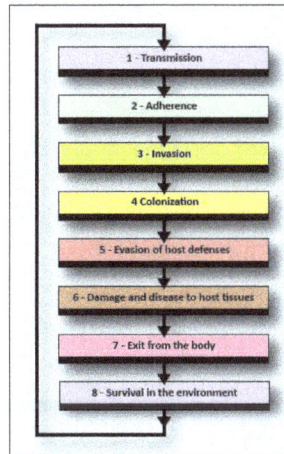

Mechanisms of pathogenesis.

Invasion

At this point, microbes begin to invade the host and produce a bacteremia (i.e., presence of bacteria in the bloodstream) or viremia (presence of a virus in the bloodstream). Microorganisms are exposed to many barriers after introduction into the host. Some bacteria are able to cause disease while remaining on the epithelial barriers, while many need to penetrate that barrier. Once this barrier has been penetrated, these pathogens can multiply without competition.

Colonization

Colonization is the multiplication of pathogenic organisms where toxins are produced and the normal flora are overcome. During this stage, pathogens Pathogens usually colonize host tissues that are in contact with the external environment. During colonization,the host begins to show signs of septicemia (i.e., blood infection where bacteria are reproducing). For infection to proceed an infectious dose should be determined. This is the minimal number of microbes necessary to establish infection. Certain pathogens are less contagious and therefore require larger numbers of pathogens to cause disease (i.e., 10-100 for Shigella and 1,000,000 for Salmonella).

Evasion of Host Defenses

After colonization, pathogens circumvent the host's innate nd adapted defenses by phagocytosis. Multiple mechanisms are used by pathogens to evade a host's immune system. For the innate system this includes:

- Intracellular pathogens that live inside a host cell

- Avoid phagocyte recognition by producing capsules prevents phagocytosis

- Producing membrane damaging toxins which can kill phagocytes (e.g., leukocidins)

- Interfere with complement activation

- Survive in the phagocyte

Pathogens must also avoid adapted defenses. Pathogens can produce proteases (i.e., allow each pathogen to avoid antibodies) or catalases (i.e., prevents the digestion of an engulfed pathogen). They can also utilize antigenic variation to alter the antigen structure. In addition, pathogens can mimic host molecules, which can cause disease-related damage.

Cause Damage or Disease to Host While Avoiding Host Defenses

Damage can occur through direct or indirect pathways. Direct methods produce toxins, which are poisonous substances that produce toxemia within a host.

Three types of toxins are produced to cause damage:

- Exotoxins: Proteins secreted by pathogens that cause damage to the host (botulinum toxin, tetanus toxin, hemolysin (ruptures red blood cells).

- Endotoxins: Toxic substances that are released when a cell is killed (Lipolysaccharides).

- Exoenzymes: Enzymes that function outside the host cells or tissues. These include co-agulase (forms a fibrin clot that "hides" the microbe from phagocytosis), hyaluronidase (breaks connective tissues down), or fibrinase (breaks down blood clots to allow pathogens to continue spreading).

Exiting the Host

A pathogen must exit the body. This occurs through various routes. Examples include sneezing, coughing, diarrhea, coitus, pus, blood, or insect bites.

Survival Outside the Host

Finally, a pathogen must be able to survive in the environment long enough to be transmitted to another host. Some are hardy and can survive for several weeks before a new host is found. There are others that survive in animal reservoirs or require direct contact because they are fragile.

Mosquito.

Pathogens may exit the body through a cough or sneeze.

Infectious Diseases: The Basics

An infectious disease is a clinically evident deviation from health. It occurs when there is a parasitic relationship between a host and a microorganism. Several different factors influence a microorganism's relationship to its host and level of severity. These include:

- Pathogenicity: The ability to produce disease in a host organism.

- Virulence: The degree of pathogenicity of a microorganism. Determinants of virulence for a pathogen include a pathogen's genetic, biochemical,or structural features. For example, one strain of influenza may only cause a fever and sore throat,while another may cause pneumonia or other serious respiratory condition.

- Infectivity: The level at which a microorganism is able to infect or invade a host.

- Transmissibility: The measure of a microorganism's ability to spread from one host to the next. This can include both distance and number of affected individuals.

Disease Stages

Infections can be either subclicinical or inapparent. If clinical signs are seen, disease occurs. The incubation period is the time between infection and first appearance of signs or symptoms. This period depends on the dose level of microbes, microbe type, virulence, and host health.

The prodromal period of 1 to 2 days follows incubation in some diseases. This includes early, mild signs/ symptoms such as, fatigue, muscle aches, and/or headache. Recovery occurs when signs/ symptoms are subsiding but the host may remain more susceptible to secondary, opportunistic infections. During the convalescent period, the body returns to a pre-diseased state.

Transmission Methods

Contact Transmission

Contact transmission includes direct contact or indirect contact. Person-to-person transmission is a form of direct contact transmission. Here the agent is transmitted by physical contact between two individuals through actions such as touching, kissing, sexual intercourse, or droplet sprays. Direct contact can be categorized as vertical, horizontal, or droplet transmission. Vertical direct contact transmission occurs when pathogens are transmitted from mother to child during pregnancy, birth, or breastfeeding. Other kinds of direct contact transmission are called horizontal direct contact transmission. Often, contact between mucous membranes is required for entry of the pathogen into the new host, although skin-to-skin contact can lead to mucous membrane contact if the new host subsequently touches a mucous membrane. Contact transmission may also be site-specific; for example, some diseases can be transmitted by sexual contact but not by other forms of contact.

Figure: Direct contact transmission of pathogens can occur through physical contact. Many pathogens require contact with a mucous membrane to enter the body, but the host may transfer the pathogen from another point of contact (e.g., hand) to a mucous membrane (e.g., mouth or eye).

When an individual coughs or sneezes, small droplets of mucus that may contain pathogens are ejected. This leads to direct droplet transmission, which refers to droplet transmission of a pathogen to a new host over distances of one meter or less. A wide variety of diseases are transmitted by droplets, including influenza and many forms of pneumonia. Transmission over distances greater than one meter is called airborne transmission.

Indirect contact transmission involves inanimate objects called fomites that become contaminated by pathogens from an infected individual or reservoir. For example, an individual with the common cold may sneeze, causing droplets to land on a fomite such as a tablecloth or carpet, or the individual may wipe her nose and then transfer mucus to a fomite such as a doorknob or towel. Transmission occurs indirectly when a new susceptible host later touches the fomite and transfers the contaminated material to a susceptible portal of entry. Fomites can also include objects used in clinical settings that are not properly sterilized, such as syringes, needles, catheters, and surgical equipment. Pathogens transmitted indirectly via such fomites are a major cause of healthcare-associated infections.

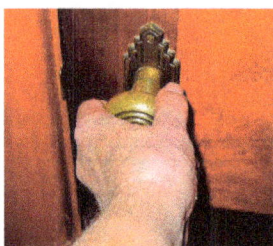

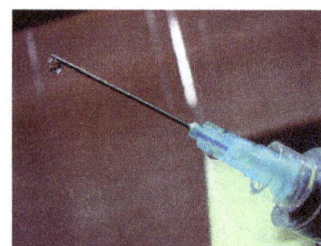

Figure: Fomites are nonliving objects that facilitate the indirect transmission of pathogens. Contaminated doorknobs, towels, and syringes are all common examples of fomites.

Vehicle Transmission

The term vehicle transmission refers to the transmission of pathogens through vehicles such as water, food, and air. Water contamination through poor sanitation methods leads to waterborne transmission of disease. Waterborne disease remains a serious problem in many regions throughout the world. The World Health Organization (WHO) estimates that contaminated drinking water is responsible for more than 500,000 deaths each year. Similarly, food contaminated through poor handling or storage can lead to foodborne transmission of disease.

Dust and fine particles known as aerosols, which can float in the air, can carry pathogens and facilitate the airborne transmission of disease. For example, dust particles are the dominant mode of transmission of hantavirus to humans. Hantavirus is found in mouse feces, urine, and saliva, but when these substances dry, they can disintegrate into fine particles that can become airborne when disturbed; inhalation of these particles can lead to a serious and sometimes fatal respiratory infection.

Although droplet transmission over short distances is considered contact transmission as discussed above, longer distance transmission of droplets through the air is considered vehicle transmission. Unlike larger particles that drop quickly out of the air column, fine mucus droplets produced by coughs or sneezes can remain suspended for long periods of time, traveling considerable distances. In certain conditions, droplets desiccate quickly to produce a droplet nucleus that is capable of transmitting pathogens; air temperature and humidity can have an impact on effectiveness of airborne transmission.

Figure: Food is an important vehicle of transmission for pathogens, especially of the gastrointestinal and upper respiratory systems. Notice the glass shield above the food trays, designed to prevent pathogens ejected in coughs and sneezes from entering the food.

Tuberculosis is often transmitted via airborne transmission when the causative agent, Mycobacterium tuberculosis, is released in small particles with coughs. Because tuberculosis requires as few as 10 microbes to initiate a new infection, patients with tuberculosis must be treated in rooms equipped with special ventilation, and anyone entering the room should wear a mask.

Clinical Focus: Florida, Resolution

This example continues the story that started on The Language of Epidemiologists and Tracking Infectious Diseases.

After identifying the source of the contaminated turduckens, the Florida public health office notified the CDC, which requested an expedited inspection of the facility by state inspectors. Inspectors found that a machine used to process the chicken was contaminated with Salmonella as a result of substandard cleaning protocols. Inspectors also found that the process of stuffing and packaging the turduckens prior to refrigeration allowed the meat to remain at temperatures conducive to bacterial growth for too long. The contamination and the delayed refrigeration led to vehicle (food) transmission of the bacteria in turduckens.

Based on these findings, the plant was shut down for a full and thorough decontamination. All turduckens produced in the plant were recalled and pulled from store shelves ahead of the December holiday season, preventing further outbreaks.

Vector Transmission

Diseases can also be transmitted by a mechanical or biological vector, an animal (typically an arthropod) that carries the disease from one host to another. Mechanical transmission is facilitated by a mechanical vector, an animal that carries a pathogen from one host to another without being infected itself. For example, a fly may land on fecal matter and later transmit bacteria from the feces to food that it lands on; a human eating the food may then become infected by the bacteria, resulting in a case of diarrhea or dysentery.

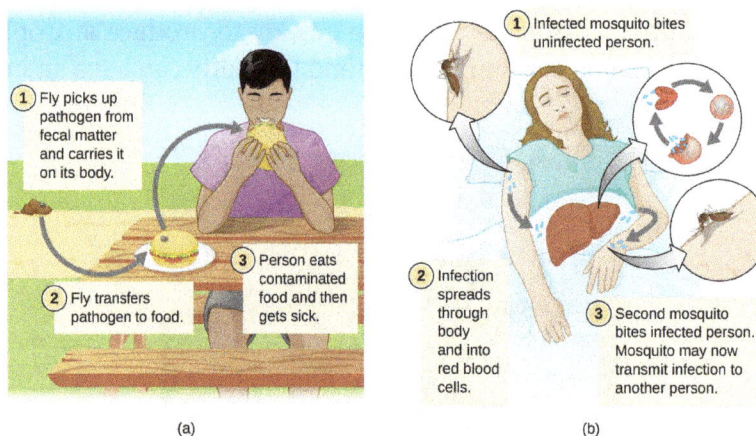

Figure (a): A mechanical vector carries a pathogen on its body from one host to another, not as an infection. (b) A biological vector carries a pathogen from one host to another after becoming infected itself.

Biological transmission occurs when the pathogen reproduces within a biological vector that transmits the pathogen from one host to another. Arthropods are the main vectors responsible for biological transmission. Most arthropod vectors transmit the pathogen by biting the host, creating a wound that serves as a portal of entry. The pathogen may go through part of its reproductive cycle in the gut or salivary glands of the arthropod to facilitate its transmission through the bite. For example, hemipterans (called "kissing bugs" or "assassin bugs") transmit Chagas disease to humans by defecating when they bite, after which the human scratches or rubs the infected feces into a mucous membrane or break in the skin.

Biological insect vectors include mosquitoes, which transmit malaria and other diseases, and lice, which transmit typhus. Other arthropod vectors can include arachnids, primarily ticks, which transmit Lyme disease and other diseases, and mites, which transmit scrub typhus and rickettsial pox. Biological transmission, because it involves survival and reproduction within a parasitized vector, complicates the biology of the pathogen and its transmission. There are also important non-arthropod vectors of disease, including mammals and birds. Various species of mammals can transmit rabies to humans, usually by means of a bite that transmits the rabies virus. Chickens and other domestic poultry can transmit avian influenza to humans through direct or indirect contact with avian influenza virus A shed in the birds' saliva, mucous, and feces.

Bacterial Infection Treatment Methods

Our bodies are a host for thousands of bacteria, and most of the time these are imperative in maintaining our bodily health. Infections as a result of bad bacteria can vary in severity from mild to fatal. These bacterial infections can be caused when harmful bacteria are exposed to your system or when bacteria reproduce haphazardly and attack other parts of your body. It is critical that you are aware of how to best treat infections quickly and effectively.

Methods to Treat a Bacterial Infection

1. Full Dosage of Antibiotics

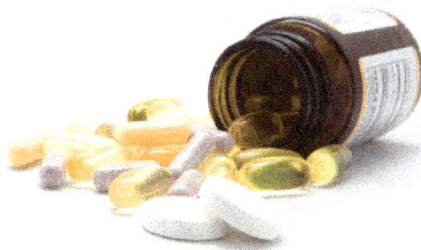

The most common treatment for bacterial infection is antibiotics. This common practice is only effective upon completion of the directed antibiotic course. Negatives effects may occur if you do not finish the recommended dosage, which include worsened infection or resistance to the antibiotic resulting in ineffective treatments in the future.

The disease and bacteria which caused the infection may be lingering in your system even after you begin to feel better. If you stop treatment prematurely, the infection may remain dormant in your system.

2. Probiotics

Probiotics have the power to help your body increase the total good bacteria while decreasing the activity from bad bacteria. When the body has more active good bacteria, it will be armed to fight off a bacterial infection. Your immune system will be stronger as a result of consumption of probiotics. Probiotics are key for treating urinary tract infections, bacterial vaginosis, bacterial skin infections, stomach or intestinal infections.

One easy way is to eat at least two cups of fresh yogurt because this contains live cultures that help manage bad bacteria. You can also find probiotic supplements.

3. Aloe Vera

There are many different methods to treat a bacterial infection naturally. Aloe Vera is a natural anti-inflammatory, high in antibacterial properties, and can aid in strengthening the body's immunity. Aloe Vera is derived from a plant and used in the form of a gel; it can treat skin, vaginal, and urinary tract infections. By consuming Aloe Vera juice in addition to the application of the gel, you can self-treat many bacterial skin infections.

4. Turmeric

As one of the most powerful herbs, turmeric not only has antibacterial and anti-inflammatory properties but it can even reduce and treat cancerous tumors. You can apply turmeric directly to the skin to treat a bacterial infection.

Another method of consumption is with a glass of warm milk as this will help with respiratory infections. Generally, by adding this power herb to your diet you will help your body be more preventative for bacterial infections.

5. Apple Cider Vinegar

Apple cider vinegar helps to maintain or correct the pH balance or imbalance within the area of a bacterial infection. With its anti-bacterial and anti-inflammatory elements, apple cider vinegar can prevent an infection from spreading or prevent the bad bacteria from being carried to the infection. Apply directly to the skin for topical use or ingest a teaspoon with a cup of water.

6. Cranberry Juice

Specifically for treating urinary tract or vaginal infections, cranberry juice is one of the safest, natural methods to treat a bacterial infection even for women who are pregnant. You can drink fresh, unsweetened cranberry juice throughout the day to combat dangerous bacteria from the inside.

7. Tea Tree Oil

Skin and vaginal infections triggered by bacteria can be quickly treated with tea tree oil by applying directly to the skin. It is commonly known as an anti-viral cure but it has strong anti-bacterial properties. Use caution when applying directly to the skin as it may cause a burning sensation, so dilute if necessary. With no dangerous side effects and high potency, tea tree oil can be used to aid even chronic infections.

8. Garlic

A basic staple in almost every kitchen, garlic is a versatile home remedy in that it can treat a variety of fungal and bacterial infections. While you can take garlic in the form of capsules, chewing and swallowing garlic directly is the most effective way to treat bacterial infections and support your digestive system. Garlic tea is another alternative, but generally you should eat 4-5 cloves per day in order to treat bacterial infections.

9. Ginger

One of the key methods to treat a bacterial infection is to reduce the amount of bad bacteria in the body. This can be done by consuming ginger, which will also aid in cooling your body and boosting blood circulation. Ginger tea, when you drink it 3-4 times per day, will treat a stomach infection or respiratory infection. To reduce the pain from an infection, use ginger with oils to massage the irritated skin.

10. Honey

By drinking natural, organic honey with a glass of warm water, you can drastically reduce symptoms of a bacterial infection. Honey will also protect a cut from getting infected with bad bacteria, and help in the treatment of a respiratory infection or skin infection.

11. Baking Soda

Baking soda helps largely to regulate the pH balance of your skin and the body in general. By mixing half teaspoon of baking soda with a glass of water, you can help to fight respiratory and stomach infections. A tub filled with warm water and a cup of baking soda is a way to soak the body when needing to heal a skin infection from bacteria.

12. Lemon

Another natural ingredient to fight an infection as a result of bacteria, lemon will help with respiratory infections in particular. By eliminating the accumulated mucus in respiratory tract that is produced from the infection, lemon will clear out the bacteria present in the mucus.

Chapter 7

Diverse uses of Bacteria

The diverse uses on bacteria lie in the production of foods, such as cheese, yoghurt and vinegar, production of drugs and vitamins, in genetic engineering and biotechnology, agriculture, biological control of pests, etc. The diverse uses of bacteria in food processing, genetic engineering, pest control, etc. have been covered in this chapter.

Food Processing

The most common usage for bacteria in food preparation is with dairy fermentations. Yogurt and cheeses have been made for centuries using bacteria. The ancients may not have known exactly what kind of bacteria that was needed or if what were needed was, indeed, bacteria. All they knew was that the previous batch was required to make a new one. Many people lack the ability to break down and absorb lactose, the sugar molecule in milk. As a result, it enters the gut, producing acid and gas, causing pain and diarrhea. Fermented milk products metabolize lactose into lactic acid, which is more tolerable for many people. The most common fermented milk product is yogurt. The lactobacilli used in the making of many yogurts, however, may not be the same type as found within the common flora of humans as there are many different strains The following are some of the bacteria used in the dairy industry:

- Acidophilus milk is made with Lactobacillus acidophilus.

- Butter is made from pasteurized cream, to which a lactic acid starter has been added. The starter contains, for example, Streptococcus cremoris or S. lactis, but requires Lactobacillus diacetylactis to give it its characteristic flavor and odor.

- Cheese is often made with Streptococcus and Lactobacillus bacteria. Fermentation lowers the pH, thus helping in the initial coagulation of the milk protein, as well as giving characteristic flavors. In such Swiss cheeses as Emmentaler and Gruyere, the typical flavor is the result of the use of Propionibacterium. Cheese can be classified within two groups - ripened and unripened. Unripened cheeses consist of cottage cheese, cream cheese, and Mozzarella, for example. These are soft cheeses and are made by the lactic acid fermentation of milk. Many different bacteria are used to produce the various cheeses, but Lactococcus lactis and Leuconostoc cremoris are used most often. Soft cheeses can take one to five months to ripen; hard cheeses, three months to a year or more; and very hard cheeses, like Parmesan, can take twelve to eighteen months. The blue veins found in cheeses, like Stilton and Roquefort, are caused by growth Penicillium roqueforti, which is deliberately added now to cheese. Originally, it was found as a natural contaminant of the areas where it was made. The holes in Swiss cheese are the result of Propionibacterium shermanii. The surfaces of Camembert and Brie are innoculated with Penicillium camembertii, which then develops in a skin on the surface. Limburger is soaked in brine to encourage the growth of Brevibacterium linens.

- Kefir includes many different microbes, including yeasts, lactobacilli, lactococci, and leuconostocs. Depending on geographical locations, the precise types of microbes will vary.

- Yogurt usually requires the addition of Lactobacillus bulgaricus, Lactococcus thermophilus, and/or Streptococcus thermophilus to the milk.

Bacteria are not only used for fermentaion in the dairy industry, but for use in other such food production as in the processing of coffee and cocoa, the manufacturing of food additives, and other such processes such as the making of xanthan gum and vinegar. Bacteria, and most viruses do not tolerate acids. This is the reason that vinegar retards the growth of most bacteria. The following is a list of other foods where bacteria and other microbes are necessary for the making of certain foodstuffs.

- Glutamic acid requires Corynebacterium glutamicum for its formation. Biotin is a cofactor essential for lipid synthesis in bacteria. By growing C. glutamicum on limited amounts of biotin, it causes the bacterial membrane to leak sufficient quantities of glutamic acid.

- Lysine -- The bacterium, Brevibacterium flavum is used in the industrial biosynthesis of lysine. Mutants no longer susceptible to feedback inhibition have been isolated to be used industrially to increase the yield of amino acids.

- Baker's yeast (Saccharomyces cerevisiae) provides a variety of enzymes that enable carbohydrates to be broken down producing sufficient carbon dioxide to give bread its characteristic texture.

- Beers, etc. -- Traditionally, the natural yeasts on grape skins determine the quality of wine produced. These natural yeasts, especially Saccharomyces cerevisiae (beer in Spanish is "cervesa") and Saccharomyces ellipsoideus, ferment the grapes to make wine. It used to be a risky business leaving it up to nature to decide the quality. Now winemaking has become a regulated science with the use of these yeasts. Beers, lagers, and ales generally rely on the yeast Saccharomyces cerevisiae, although lager yeasts will probably always be known as Saccaromyces carlsbergensis.

- Sauerkraut-making requires the bacteria Leuconostoc mesenteroides and Lactobacillus brevis to ferment sugars that provide a variety of such organic products as lactic acid, acetic acid, ethanol, and mannitol. These bacteria are known as 'heterofermentative' bacteria. Later a 'homofermentative' bacteria, Lactobacillus plantarum takes over, producing only lactic acid. Later, Enterococcus faecalis and Pediococcus cerevisiae assume the fermentation process if the salt brine is not what it sould be.

- Sourdough bread requires the help of a yeast, Saccharomyces exiguus, along with lactobacilli, to provide its characteristic texture and flavor.

- Dill pickles are simply fermented cucumbers. Streptococci starts the process of fermentation, but as the pH level falls, leuconostoc and pediococcus species, as well as Lactobacillus plantarum continue the process.

- Olives are edible only after fermentation with Lactobacillus plantarum and Lactobacillus mesenteroides.

Coffee and chocolate require Erwinia dissolvens, leuconostoc, and lactobacillus species plus the yeasts of the genus Saccharomyces to remove the tough outer coats. The microbes do not affect the taste of coffee but are necessary to confer the characteristic taste to cocoa and chocolate. The bacteria S. napoli and S. eastbourne often use chocolate as a vector. It is thought that the chocolate provides protection for the bacterium as it passes through the acidic environment of the stomach. This was observed when higher incidents of illness were reported in children.

Soy sauce is made from a mixture of soy beans and rice fermented by a variety of bacteria and fungi. These include Lactobacillus delbrueckii, Aspergillus oryzae, Aspergillus soyae, and Saccharomyces rouxii.

- Meat products, like salami and bologna sausages, require some fermentation with Pediuococcus cerevisiae, Lactobacillus plantarum and some members of the genus Bacillus. Country cured hams use fungi of the genus Aspergillus and the genus Penicillium in their fermentation process. Izushi (sushi), a Japanese delicacy made from a mixture of fish, rice, and other vegetables is produced by fermentation with lactobacilli.

Biotechnology

The biotechnology industry uses bacterial cells for the production of biological substances that are useful to human existence, including fuels, foods, medicines, hormones, enzymes, proteins, and nucleic acids. The possibilities of biotechnology are endless considering the gene reservoirs and genetic capabilities within the bacteria. Pasteur said it best, "Never underestimate the power of the microbe.

Microbes and Biotechnology

Microbes/micro-organisms are mostly micropsic small creatures are placed in different groups such bacteria, fungi, protozoa, micro-algae and viruses. These organisms live in soil, water, food, animal intestines and other different environments. Various microbial habitats reflect an enormous diversity of biochemical and metabolic traits that have arisen by genetic variation and natural selection in microbial populations.

Men used some of microbial diversity in the production of fermented foods such as bread, yogurt, and cheese. Some soil microbes release nitrogen that plants need for growth and emit gases that maintain the critical composition of the Earth's atmosphere.

Other microbes challenge the food supply by causing yield-reducing diseases in food-producing plants and animals. In our bodies, different microbes help to digest food, ward off invasive organisms, and engage in skirmishes and pitched battles with the human immune system in the give-and-take of the natural disease process.

A genome is the totality of genetic material in the DNA of a particular organism. Genomes differ greatly in size and sequence across different organisms. Obtaining the complete genome sequence of a microbe provides crucial information about its biology, but it is only the first step toward understanding a microbe's biological capabilities and modifying them, if needed, for agricultural purposes.

Microbial biotechnology, enabled by genome studies, will lead to breakthroughs such as improved vaccines and better disease-diagnostic tools, improved microbial agents for biological control of plant and animal pests, modifications of plant and animal pathogens for reduced virulence, development of new industrial catalysts and fermentation organisms, and development of new microbial agents for bioremediation of soil and water contaminated by agricultural runoff.

Microbial biotechnology is an important area that promotes for advances in food safety, food security, value-added products, human nutrition and functional foods, plant and animal protection, and overall fundamental research in the agricultural sciences.

Microbial Biotechnology and its Applications in Agriculture

Natural Fermentation

Micro-organisms found in the soil to improve agricultural productivity. Men use naturally occurring organisms to develop biofertilizers and bio-pesticides to assist plant growth and control weeds, pests, and diseases.

Micro-organisms that live in the soil actually help plants to absorb more nutrients. Plants and these friendly microbes are involved in "nutrient recycling". The microbes help the plant to "take up" essential energy sources. In return, plants donate their waste by-products for the microbes to use for food. Scientists use these friendly micro-organisms to develop biofertilizers.

Biofertilizers

Phosphate and nitrogen are important for the growth of plants. These compounds exist naturally in the environment but plants have a limited ability to extract them. Phosphate plays an important role in crop stress tolerance, maturity, quality and directly or indirectly, in nitrogen fixation. A fungus, Penicillium bilaii helps to unlock phosphate from the soil. It makes an organic acid, which dissolves the phosphate in the soil so that the roots can use it. Biofertilizer made from this organism is applied by either coating seeds with the fungus as inoculation, or putting it directly into the ground. Rhizobium is a bacteria used to make biofertilizers. This bacterium lives on the plant's roots in cell collections called nodules. The nodules are biological factories that can take nitrogen out of the air and convert it into an organic form that the plant can use.

This fertilization method has been designed by nature. With a large population of the friendly bacteria on its roots, the legume can use naturally-occurring nitrogen instead of the expensive traditional nitrogen fertilizer.

Biofertilizers help plants use all of the food available in the soil and air, thus allowing farmers to reduce the amount of chemical fertilizers they use. This helps preserve the environment for the generations to come.

Bio-pesticides

Microorganisms found in the soil are all not so friendly to plants. These pathogens can cause disease or damage the plant. As scientists developed biological "tools," which use these disease-causing microbes to control weeds and pests naturally.

Bio-herbicides

Weeds are the problem for farmers. They not only compete with crops for water, nutrients, sunlight, and space but also harbor insect and disease pests; clog irrigation and drainage systems; undermine crop quality; and deposit weed seeds into crop harvests.

Bio-herbicides are another way of controlling weeds without environmental hazards posed by synthetic herbicides. The microbes possess invasive genes that can attack the defence genes of the weeds, thereby killing it.

The benefit of using bioherbicides is that it can survive in the environment long enough for the next growing season where there will be more weeds to infect. It is cheaper than synthetic pesticides thus could essentially reduce farming expenses if managed properly. Further, it is not harmful to the environment compared to conventional herbicides and will not affect non-target organisms.

Bioinsecticides

Biotechnology can also help in developing alternative controls to synthetic insecticides to fight against insect pests. Micro-organisms in the soil that will attack fungi, viruses or bacteria, which cause root diseases. Formulas for coatings on the seed (inoculants) which carry these beneficial organisms can be developed to protect the plant during the critical seedling stage.

Bioinsecticides do not persist long in the environment and have shorter shelf lives; they are effective in small quantities, safer to humans and animals compared to synthetic insecticides; they are very specific, often affecting only a single species of insect and have a very specific mode of action; slow in action and the timing of their application is relatively critical.

Fungal- bioinsecticides

Fungi cause diseases in some 200 different insects and this disease producing traits of fungi is being used as bioinsecticides.

Fermentation technology is used to mass production of fungi. Spores are harvested and packaged so these are applied to insect-ridden fields. When the spores are applied, they use enzymes to break through the outer surface of the insects' bodies. Once inside, they begin to grow and eventually cause death.

Fungal agents are recommended by some researchers as having the best potential for long-term insect control. This is because these bioinsecticides attack in a variety of ways at once, making it very difficult for insects to develop resistance.

Virus-based Bioinsecticides

Baculoviruses affect insect pests like corn borers, potato beetles, flea beetles and aphids. One particular strain is being used as a control agent for bertha army worms, which attack canola, flax, and vegetable crops. Traditional insecticides do not affect the worm until after it has reached this stage and by then much of the damage has been done.

Microbiology Ecology Biotechnology and Sustainable Agriculture

Now increasing attention has been paid to the development of sustainable agriculture in which the high productivities of plants and animals are ensured using their natural adaptive potentials, with a minimal disturbance of the environment. It is our view that the most promising strategy to reach this goal is to substitute hazardous agrochemicals (mineral fertilizers, pesticides) with environment-friendly preparations of symbiotic microbes, which could improve the nutrition of crops and livestock, as well as their protection from biotic (pathogens, pests) and abiotic (including pollution and climatic change) stresses.

Therefore, agricultural microbiology is the present paramount research field responsible for the transfer of knowledge from general microbiology and microbial ecology to the agricultural biotechnologies. The present review is focussed on plants, but also emphasises the importance of micro-organisms in relation to agriculture and environmental health and to the biocontrol of phytophagans.

The broad application of microbes in sustainable agriculture is due to the genetic dependency of plants on the beneficial functions provided by symbiotic cohabitants. The agronomic potential of plant–microbial symbioses proceeds from the analysis of their ecological impacts, which have been best studied for N_2 fixing. This analysis has been based on 'applied co-evolutionary research' addressing the ecological and molecular mechanisms for mutual adaptation and parallel speciation of plant and microbial partners. For plant–fungal interactions, it has been demonstrated that the host genotype represents the leading factor in the biogeographic distribution of mycobionts and for their evolution within the mutualist↔antagonist and specialist↔generalist continua.

The major impact of agricultural microbiology on sustainable agriculture would be to substitute agrochemicals (mineral fertilizers, pesticides) with microbial preparations. However, this substitution is usually partial and only sometimes may be complete, e.g. in recently domesticated leguminous crops, which retain a high potential for symbiotrophic N nutrition, typical for many wild legumes. The application of nutritional symbionts could be based on plant mixotrophy, e.g. on a simultaneous symbiotrophic and combined N nutrition. This is why the maximal productivity of the majority of crops is reached using an optimal (species- and genotype-specific) combination of both nutritional types because of which a high sustainability of legume production may be achieved. Moreover, the energy costs for N_2 fixation and for assimilation of combined N differ by less than 10%. The balance between symbiotrophic and combined N nutrition may be improved by a rapid removal of N-compounds from the actively N2-fixing symbioses, as has been suggested for tropical forest ecosystems.

This approach is most promising in legume–rhizobia symbioses where the strong correlations between the ecological efficiency of mutualism and its genotypic specificity are evident.

At present, a wide spectrum of preparations of diverse microbial species may be used to enhance crop production. However, different approaches for improving the nutritional and defensive types of microbial mutualists need to be developed. For the nutritional types, an effective colonisation of plants in a host-specific manner is optimal and the impacts of beneficial symbionts are increased in parallel with their host specificity.

The application of microbial symbiotic signals or their derivatives for remodelling plant developmental or defensive functions may represent a promising field for agricultural biotechnology.

The prospects for a future development of agricultural microbiology may involve the construction of novel multipartite endo- and ecto-symbiotic communities based on extended genetic and molecular (metagenomic) analyses. The primary approach for such construction is to create composite inoculants, which simulate the natural plant-associated microbial communities. For balancing the host plant metabolism, a combination of N- and P-providing symbionts would appear promising, including the endosymbiotic rhizobia + VAM-fungi.

The further development of agricultural microbiology faces several important ecological and genetic challenges imposed by the broad application of symbiotic microbes. Some of these challenges are associated with opportunistic or even regular human pathogens, which are frequently found in endophytic communities, including Bacillus, Burkholderia, Enterobacter, Escherichia, Klebsiella, Salmonella and Staphylococcus species.

An increased knowledge of microbe-based symbioses in plants could provide effective ways of developing sustainable agriculture in order to ensure human and animal food production with a minimal disturbance of the environment. The effective management of symbiotic microbial communities is possible using molecular approaches based on the continuity of microbial pools which are circulating regularly between soil-, plant- and animal-provided niches in natural and agricultural ecosystems. Analysis of this circulation could enable the creation of highly productive microbe-based sustainable agricultural system, whilst addressing the ecological and genetic consequences of the broad application of microbes in agricultural practice.

Environmental Health and Microbial Biotechnology

Institute at ASU, addressed the challenges and solution of environmental health by manipulation of microbes.

Their solution: a synergistic marriage of two distinct disciplines, microbial ecology and environmental biotechnology. "Together, they offer much promise for helping society deal with some of its greatest challenges in environmental quality, sustainability, security, and human health," Rittmann stated in an excerpt from the paper.

Leading the marriage are revolutionary changes in compiling vast amounts of genetic information on microbial organisms through state-of-the-art DNA-based techniques. Identifying just a single microbial specimen is a daunting task, considering, that there may be trillions of bacteria in every litre of water.

To aid in the identification and function of individual micro-organisms and communities, the first use of modern molecular biology tools began in the early 1980s, with the advent of polymerase chain reaction (PCR) amplification of microbial DNA and a new view of the evolution of organisms based on their ribosomal RNA.

These technologies have advanced into high-throughput genomic and proteomic protocols that can detect specific genes and their metabolic functions with great precision and detail. Other methods can now reconstruct entire genomes of what were once "uncultivable" microbes.

With recent advances in biology, materials, computing, and engineering, environmental biotechnologists now are able to use microbial communities for a wealth of services to society.

These include detoxifying contaminated water, wastewater, sludge, sediment, or soil; capturing renewable energy from biomass; sensing contaminants or pathogens; and protecting the public from dangerous exposure to pathogens.

"Scientifically, it might be easiest to let the microbes convert the energy is organic wastes directly to electricity. However, they also can generate useful fuels, such as methane and hydrogen, and we are pursuing research on all of these renewable-energy forms."

Rittman hoped the success in capturing the energy out of waste materials, this would be a world-transforming technology and a real step forward to using more renewable forms of energy and much less reliance on fossil fuel."

Digestion

The human gut contains tens of trillions of bacteria. In fact, there are ten times more microbial cells than human cells on and in your body. The microbes in your gut are quite diverse. There are more than 1,000 different species of bacteria living in your gut alone, although only 150 to 170 species have significant populations there.

Some of these species are harmful pathogens that can cause illness. Other bacteria are helpful and provide a variety of health benefits. Bacteria living in the gut play an important role in digestion, such as helping your body break down food and absorb nutrients.

Feeding your Gut Bacteria Helps Improve your Digestion

When you eat, you are also feeding the bacteria living in your gut. Just like you, these bacteria love to feast on carbohydrates, milk sugars and proteins. Both you and the microbes living in your gut benefit from this symbiotic dining relationship.

Carbohydrates arace important energy-providing nutrients for you and your gut microbes. Your body breaks down carbohydrates into glucose, a form of sugar that your body and the microbes within it use as fuel. Scientists categorize carbohydrates as either simple or complex, where a simple carbohydrate is fast and easy to digest and a complex carbohydrate takes more time.

Like other mammals, your body is well equipped to absorb simple sugars, known as monosaccharides, in the first segments of your digestive tract. Your small intestine can easily break down the sugars in milk, table sugar, and breads, for example.

Your body has a much more difficult time digesting complex carbohydrates, known as polysaccharides, such as the fiber and starch in grains, potatoes and legumes. Your small intestine has some luck breaking starches down but quite a bit of the fiber and other nutrient-rich polysaccharides pass into the lower portions of your digestive tract undigested.

Fortunately, the microbes living in your gut also get their energy from carbohydrates. The microbes living in the large intestine break down complex carbohydrates to release the nutrition in forms your body can absorb.

The microbes in your gut also help your body break down the sugars in milk. Lactose is a type of sugar found in dairy products. Cells linking the small intestine produce lactase, a type of enzyme that breaks down lactose into simpler molecules called glucose and galactose. People with lactose intolerance do not produce lactase, which means they cannot digest dairy products and experience symptoms such as diarrhea, nausea, abdominal pain and flatulence after eating milk, cheese, or other foods containing lactose.

Fortunately, certain types of bacteria can break down lactose. Two strains of beneficial bacteria living in the gut, Lactobacillus acidophilus and Lactobacillus bulgaricus, produce lactase. These strains are also available in Natren's probiotic supplements.

Bacteria in the Lactobacillales order, also known as lactic acid bacteria, can also improve the digestibility of milk products. Specifically, these bacteria may improve the digestibility of the protein casein, found in all types of mammals' milk. For some people, consuming casein can cause bloating, pain, gas, diarrhea or gastroesophageal reflux. Yogurt and fermented cheese contain lactic acid bacteria that help improve digestion of casein.

The bacteria in your gut can help your body digest other types of proteins. Some bacteria are proteolytic, which means they break down proteins into molecules your digestive tract can absorb. The bacteria Lactobacillus bulgaricus (L. bulgaricus) breaks down the protein in yogurt, for example, to make the protein from cultured yogurt twice as digestible as regular milk protein. In other words, the bacteria in yogurt can potentially double the amount of milk protein available for absorption by your body.

L. bulgaricus may also help produce small peptides and free-form amino acids that are readily absorbed into your body. The production of free-form amino acids also helps your body absorb minerals. Natren is probably the only probiotic manufacturer that sells a pure probiotic strain of L. bulgaricus available in powders, capsules and chewable wafers.

You can optimize digestion of your next meal by supporting the population of beneficial bacteria living in your gut. Probiotic supplements contain the beneficial bacteria you need to keep your digestive tract in top running order.

Genetic Engineering

Genetic engineering is the deliberate manipulation of DNA, using techniques in the laboratory to alter genes in organisms. Even if the organisms being altered are not microbes, the substances and techniques used are often taken from microbes and adapted for use in more complex organisms.

Steps in Cloning a Gene

Let us walk through the basic steps for cloning a gene, a process by which a gene of interest can be replicated many times over. Let us pretend that we are going to genetically engineer E. coli cells to glow in the dark, a characteristic that they do not naturally possess.

1. Isolate DNA of interest: First we need to identify the genes or genes that we are interested in, the target DNA. If we want our E. coli cells to glow in the dark, we need to find an organism

that possesses this trait and identify the gene or genes responsible for the trait. The green fluorescent protein (GFP) commonly used as an expression marker in molecular techniques was originally isolated from jellyfish. In cloning a gene it is helpful to use a cloning vector, typically a plasmid or virus, capable of independent replication that will stably carry the target DNA from one location to another. Plasmid vectors are available from both bacteria and yeast.

2. Cut DNA with restriction endonucleases: Once the target and vector DNA have been identified, both types of DNA are cut using restriction endonucleases. These enzymes recognize short sequences of DNA that are 4-8 bp long. The enzymes are widespread in both bacteria and archaea, with each enzyme recognizing a specific inverted repeat sequence that is palindromic (reads the same on each DNA strand, in the 5' to 3' direction).

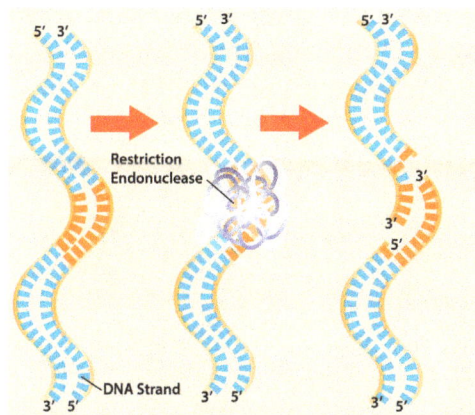

Restriction Endonucleases.

While some restriction endonucleases cut straight across the DNA (i.e. blunt cut), many make staggered cuts, producing a very short region of single-stranded DNA on each strand. These single-stranded regions are referred to as "sticky ends," and are invaluable in molecular cloning since the unpaired bases will recombine with any DNA having the complementary base sequence.

3. Combine target and vector DNA: After both types of DNA have been cleaved by the same restriction endonuclease, the two types of DNA are combined together with the addition of DNA ligase, an enzyme that repairs the covalent bonds on the sugar-phosphate backbone of the DNA. This results in the creation of recombinant DNA, DNA molecules that contain the DNA from two or more sources, also known as chimeras.

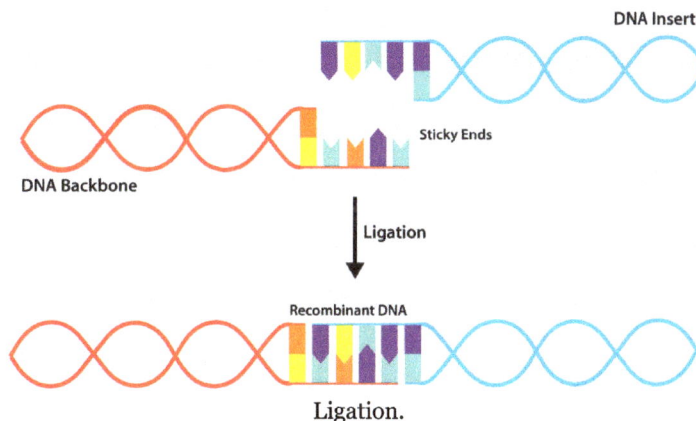

Ligation.

Introduce recombined molecule into host cell – once the target DNA has been stably combined with vector DNA, the recombinant DNA must be introduced into a host cell, in order for the genes to be replicated or expressed. There are different methods for introducing the recombinant DNA, largely depending upon the complexity of the host organism. In the case of bacteria, transformation is often the easiest method, using competent cells to pick up the recombinant DNA molecules. Alternatively, electroporation can be used, where the cells are exposed to a brief pulse of high –voltage electricity causing the plasma membrane to become temporarily permeable to DNA passage.While some cells will acquire recombinant DNA with the appropriate configuration (i.e. target DNA combined with vector DNA), the method also will yield cells carrying recombinant DNA with alternate DNA combinations (i.e. plasmid DNA combining with another plasmid DNA molecule or target DNA attached to more target DNA). The mixture is referred to as a genomic library and must be screened to select the appropriate clone. If random fragments of DNA were originally used (instead of isolation of the appropriate target DNA genes), the process is referred to as shotgun cloning and can yield thousands or tens of thousands of clones to be screened.

Introducing Recombinant DNA into Cells other than Bacteria

Agrobacterium Tumefaciens and the Ti Plasmid

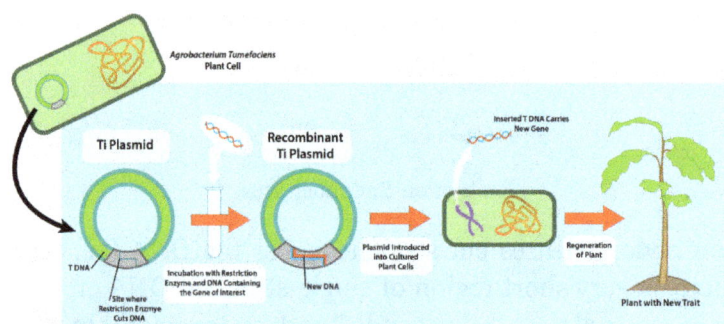

Agrobacterium Tumefaciens
Tumor-Inducing Plasmid.

Agrobacterium tumefaciens is a plant pathogen that causes tumor formation called crown gall disease. The bacterium contains a plasmid known as the Ti (tumor inducing) plasmid, which inserts bacterial DNA into the host plant genome. Scientists utilize this natural process to do genetic engineering of plants by inserting foreign DNA into the Ti plasmid and removing the genes necessary for disease, allowing for the production of transgenic plants.

Gene Gun

A gene gun uses very small metal particles (microprojectiles) coated with the recombinant DNA, which are blasted at plant or animal tissue at a high velocity. If the DNA is transformed or taken up by the cell's DNA, the genes are expressed.

Viral Vectors

For a viral vector, virulence genes from a virus can be removed and foreign DNA inserted, allowing the virus capsid to be used as a mechanism for shuttling genetic material into a plant or animal cell. Marker genes are typically added that allow for identification of the cells that took up the genes.

DNA Techniques

Gel Electrophoresis

Gel electrophoresis is a technique commonly used to separate nucleic acid fragments based on size. It can be used to identify particular fragments or to verify that a technique was successful.

A porous gel is prepared made of agarose, with the concentration adjusted based on expected size. Nucleic acid samples are deposited into wells in the gel and an electrical current is applied. Nucleic acid, with its negative charge, will move towards the positive electrode, which should be placed at the bottom of the gel. The nucleic acid will move through the gel, with the smallest pieces encountering the least resistance and thus moving through the fastest. The length of passage of each nucleic acid fragment can be compared to a DNA ladder, with fragments of known size.

Polymerase Chain Reaction (PCR)

The polymerase chain reaction or PCR is a method used to copy or amplify DNA in vitro. The process can yield a billionfold copies of a single gene within a short period of time. The template DNA is mixed with all the ingredients necessary to make DNA copies: primers (small oligonucleotides that flank the gene or genes of interest by recognizing sequences on either side of it), nucleotides (the building blocks of DNA), and DNA polymerase. The steps involve heating the template DNA in order to denature or separate the strands, dropping the temperature to allow the primers to anneal, and then heating the mixture up to allow the DNA polymerase to extend the primers, using the original DNA as an initial template. The cycle is repeated 20-30 times, exponentially increasing the amount of target DNA in a few hours.

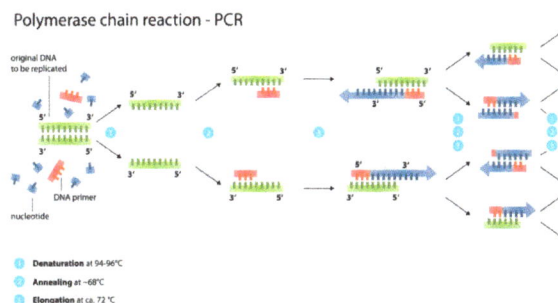

Polymerase Chain Reaction (PCR).

Uses of Genetically Engineered Organisms

There can be numerous reasons to create a genetically modified organism (GMO) or transgenic organism, defined as a genetically modified organism that contains a gene from a different organism. Typically the hope is that the GMO will provide needed information or a product of value to society.

Source of DNA

Genetically engineered organisms can be made so that a piece of DNA can be easily replicated, providing a large source of that DNA. For example, a gene associated with breast cancer can be spliced into the genome of E. coli, allowing for the rapid production of the gene so that it may be sequenced, studied, and manipulated, without requiring repeated tissue donations from human volunteers.

Source of RNA

Antisense RNA is ssRNA that is complementary to the mRNA that will code for a protein. In cells it is made as a way to control target genes. There has been increasing interest in the use of antisense RNA as a way to prevent diseases that are caused by the production of a particular protein.

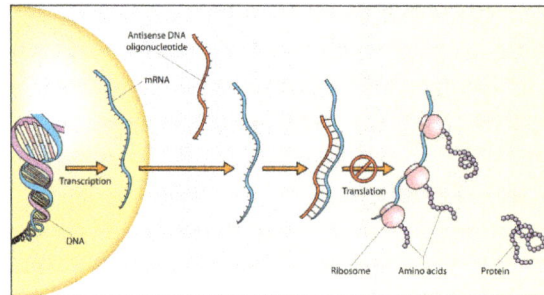

Antisense RNA.

Source of Protein

Since microbes replicate so rapidly, it can be extremely advantageous to use them to manufacture proteins of interest or value. Given the right promoters, bacteria will express genes for proteins that are not naturally found in bacteria, such as cytokine. Genetically engineered cells have been used to make a wide variety of proteins of use to humans, such as insulin or human growth hormone.

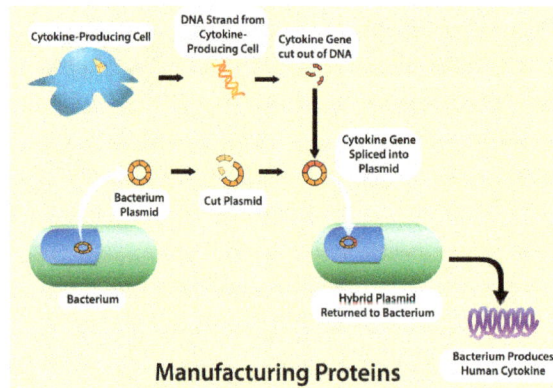

Manufacturing Proteins

Pest Control

Beneficial bacteria have been a part of horticulture since the beginning of time, though our understanding of how they work and how to use them continues to develop. Many gardeners today are familiar with beneficial microorganism products that inoculate a soil or medium.

Some of these products have specific strains of beneficial microorganisms that create synergistic relationships with the roots. These relationships aid in the breakdown of organic materials and speeds up nutrient uptake. There are other bacteria strains used in horticulture to protect plants from pest insects and pathogenic diseases.

Once used only by organic and natural horticulturists, some of these beneficial bacteria products for pest insect and disease control are now being embraced by large commercial operations. This is due to their effectiveness and the minimal negative impact most of these products have on the environment. In many cases, these beneficial bacteria products are safer and more effective than their chemical counterparts.

After all, many of the chemical pesticides and fungicides damage much more than just the pest insects or diseases they were designed to treat.

Bacteria for Insect Control

Few things are more devastating to a garden than a pest insect infestation. Pest insects can quickly destroy an otherwise healthy crop, which is why a grower must take immediate action whenever a pest insect problem is identified. Depending on the particular pest insect problem a grower faces, he or she may be able to implement a treatment with a beneficial bacteria strain.

Bacillus Thuringiensis

Bacillus thuringiensis (Bt.) is a naturally occurring, soil-dwelling bacterium that is used as a biological pesticide. It is most commonly used to treat caterpillars and is very effective at controlling cabbage loopers.

While there are many different strains of Bt. found throughout the world, some strains produce crystal proteins that have insecticidal properties. When ingested, these special crystal proteins disturb the insect's digestive system. This causes the insect to stop eating entirely, so it will eventually starve to death.

Many biological insecticidal sprays and powders contain Bt. Application instructions vary, but it is imperative that the insects consume the Bt. or the treatment will not work. In most cases, crops should be treated with Bt. on hot, dry days when the insects are most actively feeding. This will ensure a high consumption rate among the pest insect population. It is also important to remember that a reapplication may be necessary as rain or wind can remove the active ingredient from the plants.

Bacillus thuringiensis is considered an all-natural pesticide and some products containing it are certified for use by organic certification institutions. Most Bt. sprays and powders are regarded as environmentally friendly and have little or no effect on humans, wildlife, pollinators, or most beneficial insects. New Bt. strains are being developed all the time.

Spinosad (Saccharopolyspora Spinosa)

Spinosad is a biological insecticide based on chemical compounds found in the bacterial species Saccharopolyspora spinosa. The insecticide is a result from the fermentation of the bacteria, a process that creates different forms of spinosyns. Spinosyns occur in more than 20 natural forms and more than 200 synthetic forms have been created in a lab. The insecticide spinosad contains a mix of two specific spinosyns, spinosyn A and spinosyn D, in a roughly 17:3 ratio.

Unlike Bt., which must be consumed by the pest insect, spinosad is effective by both contact and ingestion. Spinosad affects receptors in the insect's nervous system, making the insect unable to feed or reproduce. Spinosad is considered an all-natural product and is approved for use in organic horticulture by numerous nations. It is used to control a wide variety of insects, including caterpillars, flies, beetles, thrips, and spider mites.

Bacteria for Fungi/Pathogen Control

Another tragedy that can befall an otherwise healthy garden is a pathogen attack. Powdery mildew, grey mold, root rot (pythium), and botrytis can all quickly destroy a garden. Similar to a pest insect problem, immediate action should be taken whenever a pathogen is identified. One of the best ways to combat these fungi-based pathogens is with the use of bacteria.

Bacillus Subtilis

Bacillus subtilis is a naturally occurring bacteria commonly found in soil and in the gastrointestinal tract of humans. It has many different uses in many different industries, but in horticulture, Bacillus subtilis is used as a natural treatment for powdery mildew.

When sprayed directly on powdery mildew, the strain feeds on the pathogenic fungus. When the fungus is gone, the bacteria dies off as well since it no longer has a food source. Unfortunately, multiple applications may be necessary as airborne powdery mildew spores can settle onto plants after the first batch of Bacillus subtilis dies off. It is also difficult to apply Bacillus subtilis in the later stage of flowering in an indoor garden because spraying any liquid will increase the level of humidity.

Bacillus Amyloliquefaciens

Bacillus amyloliquefaciens is a species of bacterium in the genus Bacillus. Bacillus amyloliquefaciens is a fast-growing rhizobacteria and can quickly colonize roots. It has gained immense popularity among hydroponic and aquaponic growers due to its ability to destroy and keep away root rot. In horticulture, Bacillus amyloliquefaciens is used to treat multiple root pathogens, including ralstonia, fusarium, and pythium.

Streptomyces Lydicus

Streptomyces lydicus is a bacterium species from the genus Streptomyces and has been isolated from soil. In horticulture, Streptomyces lydicus is used as a biological fungicide. It can be used to treat various fungal pathogens, including fusarium, pythium, phytophthora, rhizoctonia, and verticillum. Streptomyces lydicus is most commonly found in powder form that can be mixed in a liquid for application or mixed directly in the soil or medium.

The way the microbial world affects our daily lives is something most people never think about. Even experienced horticulturists take the beneficial microbes that affect a garden's performance for granted.

As the old saying goes, "out of sight, out of mind." However, as we discover more about how the microbial world affects horticulture and how we can use certain microorganisms to help us achieve our goals, we will pay closer attention to the microscopic world around us.

The link between healthy microbial life and successful gardening is unquestionable. Although we are unable to see the microorganisms, gardeners can witness the undeniable results from using beneficial microbe products in the garden.

Chapter 8

Bacterial Diseases

Bacteria can form a parasitic association with other organisms that can lead to disease and death. In humans, these are manifested as various infections such as typhoid fever, tetanus, syphilis, leprosy, etc. The aim of this chapter is to provide an overview of bacterial diseases such as cellulitis, folliculitis, impetigo, vibriosis, gonorrhea, etc.

Bacterial diseases refer to a large variety of diseases caused by bacteria or bacterial components that affect humans, domesticated animals, wildlife, fish, and birds. Most of these diseases are contagious—that is, they can be passed from one member of a species to another member, or, in a smaller number of instances, from one species to a different species. Depending on the organism, bacterial disease can be spread in different ways. Examples include contaminated food or water, air currents, infection of an environment that is not normally inhabited by the particular bacterium, and the possession or.

Cellulitis

Cellulitis is a bacterial infection of the deeper layers of the skin. It can start suddenly, and it can become serious if not treated.

If it spreads deeper into the body, it can be life-threatening.

Early treatment with antibiotics is usually successful. Most people can be treated at home, but sometimes they need to spend time in the hospital.

Cellulitis can affect any part of the body, but it is most likely to appear in the lower legs. It is a painful condition.

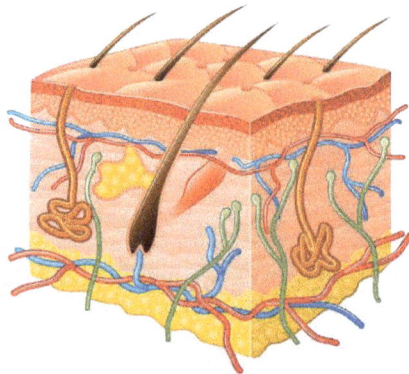

Cellulitis is an infection of the deeper layers of the skin.

Treatment

The following treatments are commonly recommended for cellulitis:

Medication

Cellulitis nearly always responds rapidly to antibiotics. Some people experience a slight worsening of the reddening of the skin at the start of antibiotic treatment, which usually subsides within a couple of days.

However, anyone who experiences fever, vomiting, or any worsening of their symptoms after starting antibiotic treatment, should contact a doctor immediately. Many different types of antibiotics can be used to treat cellulitis. Which type the doctor prescribes will depend on what type of bacteria the doctor suspects has caused the infection.

Antibiotics are normally taken for 5-10 days, but treatment might last 14 days or more in some cases.

Treatment in the Hospital

Some people with severe cellulitis may require hospital treatment, especially if the cellulitis is deteriorating, if the person has a high fever, is vomiting, fails to respond to treatment, or has recurrences of cellulitis.

Most people who are treated in hospital will receive their antibiotic through a vein in their arm (intravenously, using a drip).

Types

Cellulitis can be classified into different types, according to where it appears.

This can be:

- around the eyes, known as periorbital cellulitis
- around the eyes, nose, and cheeks, known as facial cellulitis
- breast cellulitis
- perianal cellulitis, occurring around the anal orifice

However, the most common location is the lower legs.

Symptoms

The affected area will become:

- warm
- tender
- inflamed
- swollen

- red
- painful

Some people may develop blisters, skin dimpling, or spots. They might also experience a fever, chills, nausea, and shivering.

Lymph glands may swell and become tender. If the cellulitis has affected the person's leg, the lymph glands in their groin may also be swollen or tender.

Causes

Bacteria from the Streptococci and staphylococci groups are commonly found on the surface of the skin and cause no harm, however, if they enter the skin, they can cause infection.

For the bacteria to access the deeper skin layers, they need a route in, which is usually through a break in the skin. A break in the skin can be caused by:

- ulcers
- burns
- bites
- grazes
- cuts
- some skin conditions, such as eczema, athlete's foot, or psoriasis

Some people develop cellulitis without being able to identify a break in the skin.

Risk factors

The following risk factors increase the likelihood of cellulitis.

- Leg swelling (edema): This raises the chances of developing cellulitis.
- Weakened immune system: Including people who are undergoing chemotherapy or radiotherapy, those with HIV or AIDS, and older adults.
- Diabetes: If the diabetes is not properly treated or controlled, a person's immune system can be weaker, or they may have circulatory problems, which can lead to skin ulcers.
- Blood circulation problems: People with circulation issues may develop skin infections.
- Other skin infections: Conditions, such as chicken pox and shingles may cause skin blisters. If the blisters break, they can become ideal routes for bacteria to get into the skin.
- Lymphedema: This condition causes swollen skin that is more likely to crack. Cracks in the skin may become perfect entry routes for bacteria.
- Previous cellulitis: A person who has had cellulitis before has a higher risk than others of developing it again.

- Intravenous drug users: Drug addicts who do not have access to a regular supply of clean needles are more likely suffer from infections deep inside the skin.

Diagnosis

Diagnosis is usually fairly straightforward and does not generally require any complicated tests. A doctor will examine the individual and assess their symptoms.

Most cases of cellulitis are caused by streptococci and staphylococci, but other conditions, such as Lyme disease, may look like cellulitis, so it is important to follow up with a doctor after diagnosis.

The doctor may take a swab, or sample, if there is an open wound. This can help them identify what type of bacteria is causing cellulitis.

However, these samples are easily contaminated due to the multiple types of bacteria that live on the skin all the time.

A small percentage of patients may have serious complications that include:

- Blood infection and sepsis: If the bacteria reach the bloodstream, the person has a higher risk of developing sepsis. A person with sepsis may have a fever, accelerated heartbeat, rapid breathing, low blood pressure (hypotension), dizziness when standing up, reduced urine flow, and sweaty, pale, cold skin.

- Infection moving to other regions: This is very unusual, but the bacteria that caused the cellulitis can spread to other parts of the body, including muscle, bone, or the heart valves. If this happens, the person needs treatment immediately.

- Permanent swelling: People who do not receive treatment for their cellulitis are at higher risk of having a permanent swelling in the affected area.

In most cases, cellulitis treatment is effective, and the person will not experience any complications.

Home Remedies

There is no way to treat cellulitis at home, and this condition needs to be treated by a doctor. If someone suspects they have cellulitis, they should call a doctor right away, and:

- drink plenty of water

- keep the affected area raised, to help reduce swelling and pain

- take painkillers, as recommended by a doctor

Some people have suggested using tea tree oil, coconut oil, and garlic, because they may have antibacterial, antifungal, and other properties. However, there appears to be no evidence that they can treat cellulitis.

Anyone with symptoms needs to seek medical help at once. Untreated, cellulitis can be life-threatening.

Prevention

Although some cases of cellulitis are not preventable, there are things that people can do to reduce their chances of developing it:

- Treat cuts and grazes: If the skin is broken because of a cut, bite, or graze, it should be kept clean to reduce risk of infection.

- Reduce the likelihood of scratching and infecting the skin: The risk of the skin being damaged by scratching will be greatly reduced if fingernails are kept short and clean.

- Take good care of the skin: If the skin is dry, use moisturizers to prevent skin from cracking. Individuals with greasy skin will not need to do this. Moisturizers will not help if the skin is already infected.

- Protect the skin: Wear gloves and long sleeves when gardening; do not wear shorts if there is a likelihood of grazing the skin of the legs.

- Lose weight if you are obese: Obesity may raise the risk of developing cellulitis.

Folliculitis

Folliculitis is a common skin condition in which hair follicles become inflamed. It's usually caused by a bacterial or fungal infection. At first it may look like small red bumps or white-headed pimples around hair follicles — the tiny pockets from which each hair grows. The infection can spread and turn into nonhealing, crusty sores.

The condition isn't life-threatening, but it can be itchy, sore and embarrassing. Severe infections can cause permanent hair loss and scarring.

If you have a mild case, it'll likely clear in a few days with basic self-care measures. For more serious or recurring folliculitis, you may need to see a doctor for prescription medicine.

Certain types of folliculitis are known as hot tub rash, razor bumps and barber's itch.

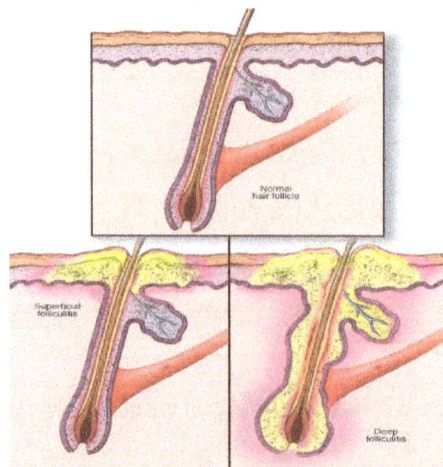

Folliculitis

Symptoms

Folliculitis signs and symptoms include:

- Clusters of small red bumps or white-headed pimples that develop around hair follicles
- Pus-filled blisters that break open and crust over
- Itchy, burning skin
- Painful, tender skin
- A large swollen bump or mass

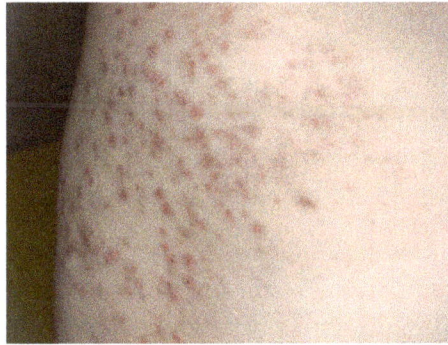

Hot Tub Folliculitis

When to See a Doctor

Make an appointment with your doctor if your condition is widespread or the signs and symptoms don't go away after a few days. You may need an antibiotic or an antifungal medication to help control the condition.

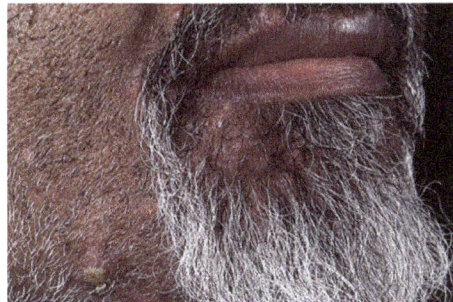

Pseudofolliculitis Barbae

Types of Folliculitis

The two main types of folliculitis are superficial and deep. The superficial type involves part of the follicle, and the deep type involves the entire follicle and is usually more severe.

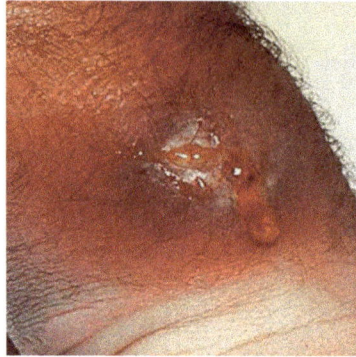

Carbuncle

Forms of superficial folliculitis include:

- Bacterial folliculitis: This common type is marked by itchy, white, pus-filled bumps. It occurs when hair follicles become infected with bacteria, usually Staphylococcus aureus (staph). Staph bacteria live on the skin all the time. But they generally cause problems only when they enter your body through a cut or other wound.

- Hot tub folliculitis (pseudomonas folliculitis): With this type you may develop a rash of red, round, itchy bumps one to two days after exposure to the bacteria that causes it. Hot tub folliculitis is caused by pseudomonas bacteria, which is found in many places, including hot tubs and heated pools in which the chlorine and pH levels aren't well-regulated.

- Razor bumps (pseudofolliculitis barbae): This is a skin irritation caused by ingrown hairs. It mainly affects men with curly hair who shave too close and is most noticeable on the face and neck. People who get bikini waxes may develop barber's itch in the groin area. This condition may leave dark raised scars (keloids).

- Pityrosporum (pit-ih-ROS-puh-rum) folliculitis: This type produces chronic, red, itchy pustules on the back and chest and sometimes on the neck, shoulders, upper arms and face. This type is caused by a yeast infection.

Forms of deep folliculitis include:

- Sycosis barbae: This type affects males who have begun to shave.

- Gram-negative folliculitis. This type sometimes develops if you're receiving long-term antibiotic therapy for acne.

- Boils (furuncles) and carbuncles: These occur when hair follicles become deeply infected with staph bacteria. A boil usually appears suddenly as a painful pink or red bump. A carbuncle is a cluster of boils.

- Eosinophilic (e-o-sin-o-FILL-ik) folliculitis: This type mainly affects people with HIV/AIDS. Signs and symptoms include intense itching and recurring patches of bumps and pimples that form near hair follicles of the face and upper body. Once healed, the affected skin may be darker than your skin was previously (hyperpigmented). The cause of eosinophilic folliculitis isn't known.

Causes

Folliculitis is most often caused by an infection of hair follicles with Staphylococcus aureus (staph) bacteria. Folliculitis may also be caused by viruses, fungi and even an inflammation from ingrown hairs.

Follicles are densest on your scalp, and they occur everywhere on your body except your palms, soles, lips and mucous membranes.

Risk Factors

Anyone can develop folliculitis. But certain factors make you more susceptible to the condition, including:

- Having a medical condition that reduces your resistance to infection, such as diabetes, chronic leukemia and HIV/AIDS
- Having acne or dermatitis
- Taking some medications, such as steroid creams or long-term antibiotic therapy for acne
- Being a male with curly hair who shaves
- Regularly wearing clothing that traps heat and sweat, such as rubber gloves or high boots
- Soaking in a hot tub that's not maintained well
- Causing damage to hair follicles by shaving, waxing or wearing tight clothing.

Complications

Possible complications of folliculitis include:

- Recurrent or spreading infection
- Boils under the skin (furunculosis)
- Permanent skin damage, such as scarring or dark spots
- Destruction of hair follicles and permanent hair loss.

Prevention

You can try to prevent folliculitis from coming back with these tips:

- Avoid tight clothes: It helps to reduce friction between your skin and clothing.
- Dry out your rubber gloves between uses. If you wear rubber gloves regularly, after each use turn them inside out, rinse with soap and water, and dry thoroughly.
- Avoid shaving, if possible: For men with razor bumps (pseudofolliculitis), growing a beard may be a good option if you don't need a clean-shaven face.
- Shave with care: If you shave, adopt habits such as the following to help control symptoms by reducing the closeness of the shave and the risk of damaging your skin:

o   Shaving less frequently.

o   Washing your skin with warm water and antibacterial soap before shaving.

o   Using a washcloth or cleansing pad in a gentle circular motion to raise embedded hairs before shaving.

o   Applying a good amount of shaving lotion before shaving.

o   Shaving in the direction of hair growth, though one study found that men who shaved against the grain had fewer skin bumps. See what works best for you.

o   Avoiding shaving too close by using an electric razor or guarded blade and by not stretching the skin.

o   Using a sharp blade and rinsing it with warm water after each stroke.

o   Applying moisturizing lotion after you shave.

o   Avoiding the sharing of razors, towels and washcloths.

- Considering hair-removing products (depilatories) or other methods of hair removal. Though they, too, may irritate the skin.

- Use only clean hot tubs and heated pools. And if you own a hot tub or a heated pool, clean it regularly and add chlorine as recommended.

- Talk with your doctor: Depending on your situation and frequency of recurrences, your doctor may suggest controlling bacterial growth in your nose with a five-day regimen of antibacterial ointment and using a body wash with chlorhexidine (Hibiclens, Hibistat.

Diagnosis

Your doctor is likely to diagnose folliculitis by looking at your skin and reviewing your medical history. He or she may use a technique for microscopic examination of the skin (dermoscopy).

If initial treatments don't clear up your infection, your doctor may use a swab to take a sample of your infected skin or hair. This is sent to a laboratory to help determine what's causing the infection. Rarely, a skin biopsy may be done to rule out other conditions.

Treatment

Treatments for folliculitis depend on the type and severity of your condition, what self-care measures you've already tried and your preferences. Options include medications and interventions such as laser hair removal. Even if treatment helps, the infection may come back.

Medications

- Creams or pills to control infection. For mild infections, your doctor may prescribe an antibiotic cream, lotion or gel. Oral antibiotics aren't routinely used for folliculitis. But for a severe or recurrent infection, your doctor may prescribe them.

- Creams, shampoos or pills to fight fungal infections. Antifungals are for infections caused by yeast rather than bacteria. Antibiotics aren't helpful in treating this type.

- Creams or pills to reduce inflammation. If you have mild eosinophilic folliculitis, your doctor may suggest you try a steroid cream to ease the itching. If you have HIV/AIDS, you may see improvement in your eosinophilic folliculitis symptoms after antiretroviral therapy.

Other Interventions

- Minor surgery: If you have a large boil or carbuncle, your doctor may make a small incision in it to drain the pus. This may relieve pain, speed recovery and lessen scarring. Your doctor may then cover the area with sterile gauze in case pus continues to drain.

- Laser hair removal: If other treatments fail, long-term hair removal with laser therapy may clear up the infection. This method is expensive and often requires several treatments. It permanently removes hair follicles, thus reducing the density of the hair in the treated area. Other possible side effects include discolored skin, scarring and blistering.

Lifestyle and Home Remedies

Mild cases of folliculitis often improve with home care. The following approaches may help relieve discomfort, speed healing and prevent an infection from spreading:

- Apply a warm, moist washcloth or compress. Do this several times a day to relieve discomfort and help the area drain, if needed. Moisten the compress with a saltwater solution (1 teaspoon of table salt in 2 cups of water).

- Apply over-the-counter antibiotics. Try various nonprescription infection-fighting gels, creams and washes.

- Apply soothing lotions. Try relieving itchy skin with a soothing lotion or an over-the-counter hydrocortisone cream.

- Clean the affected skin. Gently wash the infected skin twice a day with antibacterial soap. Use a clean washcloth and towel each time and don't share your towels or washcloths. Use hot, soapy water to wash these items. And wash clothing that has touched the affected area.

- Protect the skin. If possible, stop shaving, as most cases of barber's itch clear up a few weeks after you stop shaving.

Impetigo

Impetigo (pronounced im-puh-ty-go) is caused by a Staphylococcus aureus or Streptococcus pyogenes bacterial infection on the outer layers of skin, the epidermis. The face, arms, and legs are the skin areas most often affected.

Anyone can get impetigo, but it's the most common bacterial skin infection among children, affecting mostly 2- to 5-year olds (4, 5). In fact, it accounts for about 10 percent of skin problems seen in pediatric clinics.

The infection most often begins in minor cuts, insect bites, or a rash such as eczema — any place there is broken skin. But it can also occur on healthy skin.

It's called primary impetigo when it infects healthy skin and secondary impetigo when it occurs in broken skin.

Impetigo is an old disease. The name dates back to 14th-century England and comes from the Latin word impetere, meaning "to attack." Attack seems an appropriate name for this easily spread infection.

Contagion

The open sores are highly contagious, itchy, and sometimes painful. Scratching the sores can spread the infection from one place on your skin to another, or to another person. The infection can also spread from anything an infected person touches.

Because it spreads so easily, impetigo is also called the "school disease." It can quickly spread from child to child in a classroom or day care center where children are in close contact. For the same reason, it also spreads easily in families.

Hygiene is key to controlling impetigo's spread. If you or your child has impetigo, you need to wash and disinfect everything the infection might come in contact with, including clothes, bedding, towels, toys, or sports equipment.

Topical antibiotics can usually clear up impetigo in days, and shorten the length of time that the disease is contagious.

A Global Problem

Impetigo is a global disease that has remained at the same incidence levels for the last 45 years. An estimated 162 million children worldwide have impetigo at any one time.

Bacteria thrive in hot, moist conditions. So impetigo tends to be seasonal, peaking in the summer and fall in northern climates. But in warm and humid climates, it can occur year-round.

Impetigo is more prevalent in developing countries, and in the poor areas of industrial countries. A 2015 review of impetigo found the highest incidence in the 14 countries of Oceania. This same study recommended that more research and more attention be paid to impetigo as a public health problem.

Common Symptoms of Impetigo

Reddish spots on the skin, often clustered around the nose and lips, are the first sign of the most common type of impetigo.

The sores quickly grow into blisters, ooze and burst, and then form a yellowish crust. The crust is often described as honey-colored. The clusters of blisters may expand to cover more of your skin.

The sores are unsightly, itchy, and occasionally painful. After the crust phase, they leave red marks that fade without leaving scars.

Infants often have a less common type of impetigo, with larger blisters around the diaper area or in skin folds. These fluid-filled blisters soon burst, leaving a scaly rim called a collarette.

Impetigo can be uncomfortable. Occasionally, it may involve swollen glands in the area of the outbreak. Fever and swollen glands can occur in more severe cases.

Types of Impetigo

There are three types of impetigo distinguished by the bacteria that cause them and the sores they form.

Nonbullous

Nonbullous, also called impetigo contagiosa, is mainly caused by Staphylococcus aureus. It's the most common form of impetigo, an estimated 70 percent of cases.

Nonbullous impetigo can also be caused by Streptococcus pyogenes or by a combination of both staph and strep. A small number of cases, 5 to 10 percent, are caused by strep bacteria alone.

It usually starts with reddish spots that develop into small red blisters around the mouth and nose. The blisters range in size from 1 to 2 centimeters in diameter (.39 to .78 inch). The clusters of blisters may spread to other skin areas.

After a few days, the blisters burst and develop a brownish-yellow crust. The surrounding skin can look red and raw.

Nonbullous impetigo is itchy, but not painful. When the crusts heal, there are reddish spots that fade and don't leave scars.

Nonbullous impetigo rarely occurs in children under 2.

Bullous

Bullous impetigo is caused by Staphylococcus aureus.

It usually forms larger blisters or bullae filled with a clear fluid that becomes darker and cloudy. The blisters can be up to 2 centimeters in diameter (about .78 inch).

Typically, the blisters begin on unbroken skin and aren't surrounded by reddish areas. The blisters become limp and then burst open. Then a yellowish crust forms over the sore.

Bullous impetigo is most common in newborns, especially in the diaper area or neck folds. For other ages, the blisters appear most often on the trunk and arms and legs.

The blisters usually leave no scars when they have healed.

Ecthyma

Ecthyma is caused by Streptococcus pyogenes, Staphylococcus aureus, or both.

The infection forms small, pus-filled sores with a thicker crust. But ecthyma goes deeper into the skin than the other forms of impetigo, and it can be more severe. Ecthyma sometimes may be accompanied by swollen glands.

Ecthyma blisters can be painful and can develop into larger, deeper sores, between 0.5 and 3 centimeters in diameter (0.3 to 1.2 inches). These sores progress to have a thick crust surrounded by reddish-purple skin.

Most often ecthyma appears on your buttocks, thighs, legs, ankles, and feet. Sometimes untreated nonbullous or bullous impetigo can develop into ecthyma.

The ecthyma lesions heal slowly and may leave scars after they heal.

Causes of Impetigo

Impetigo is a bacterial infection. Your skin surface and the inside of your nose are normally home to large numbers of "friendly" or commensal bacteria that help protect you from disease-causing bacteria such as Staphylococcus aureus and Streptococcus pyogenes.

Your commensal bacteria work to keep down the population of pathogenic bacteria by producing substances that are toxic to the pathogens, depriving them of nutrients, among other measures.

But strains of these staph or strep bacteria can take advantage of a break in the skin from a cut, scratch, insect bite, or rash to invade and colonize, causing impetigo.

The bacteria can also colonize and cause an infection on normal skin. It's not known exactly why this happens.

Within about 10 days of bacteria colonization, impetigo blisters appear. The way it works is that the Staphylococcus aureus and Streptococcus pyogenes bacteria produce toxins that break apart your top skin layers, causing blisters to form.

In many cases, the bacteria are already on site, waiting for an opportunity to colonize.

Staphylococcus aureus and Streptococcus pyogenes bacteria are normally carried in the nose by between 20 and 50 percent of the general population. An even larger percentage of people are intermittent carriers.

Further, about 10 to 20 percent of healthy people have Staphylococcus aureus bacteria in their perineum (the area between the genitals and the anus).

For people who are Staphylococcus aureus carriers, infection is thought to be spread by the person from their nose or other area to the skin. In contrast, strep-caused impetigo usually begins with the strep bacteria spreading to the skin from a person with impetigo.

Normally strep doesn't survive on skin for more than a few hours. It's not known why the strep bacteria are able to stay on the skin of people who develop impetigo for 10 days, before the blisters appear.

Strains of strep bacteria behave differently. Research has shown that some strains of strep bacteria cause throat infections, while others cause skin infections.

Why do some people carry staph and strep bacteria without developing impetigo? It's thought that some individuals are more able to resist infection because of the chemical makeup of their skin and their general good health.

Other Factors in Impetigo

Other factors can make a difference in the growth of staph and strep bacteria that cause impetigo:

- Poor hygiene aids the spread of bacteria. One study found that when child caretakers had an orientation program about the importance of hand washing, the incidence of impetigo in their group was 34 percent lower.

- Disease-causing bacteria thrive in hot humid weather.

- Working or living in close crowded conditions can promote impetigo spread. This includes the military, especially in tropical areas.

- Sports that involve skin-to-skin contact, such as football, wrestling, or jiu-jitsu put you at risk.

How does Impetigo Spread

Impetigo is highly contagious. It spreads on direct contact with a skin sore or with anything that may have touched an open sore.

Though uncommon, impetigo can also spread by contact with bedding, underwear and clothes, towels and washcloths, toys, sports equipment, and anything else that came in contact with an open sore.

If you're using a topical antibiotic, the sores are contagious until they stop oozing and dry up.

If you're taking an oral antibiotic, the infection usually won't be contagious after 24 to 48 hours.

At-risk Populations

Children 2 to 5 years old, especially those in a day care center or play group, are the most at risk.

Adults and children are more at risk if they:

- live in a warm, humid climate

- have diabetes

- are undergoing dialysis (4)

- have a compromised immune system, such as from HIV

- have skin ailments such as eczema, dermatitis, or psoriasis

- have sunburn or other burns (3)

- have itchy infections such as lice, scabies, herpes simplex, or chickenpox (4)

- have insect bites or poison ivy

- participate in contact sports

When should you See a Doctor

It's a good idea to see your doctor if you suspect impetigo. Antibiotic treatment for impetigo speeds up healing and can stop the spread of infection for you (or your child) and others.

With treatment, impetigo usually heals in 7 to 10 days. If you have an underlying infection or skin disease, treatment may take longer to heal.

It's likely that your doctor can diagnose impetigo by its appearance. But in a severe case, the doctor may want to culture the bacteria.

Treatment of Impetigo

Treatment for impetigo depends on how widespread or severe the blisters are.

Antibiotics

The Infectious Diseases Society of America recommends treatment with topical antibiotics for 5 to 7 days.

The specific topical antibiotics recommended are mupirocin and fusidic acid. A 2003 meta-analysis of 16 studies found no significant difference between these two topical antibiotics.

If your impetigo is severe or widespread, oral antibiotics are recommended. These work more quickly than topical antibiotics. However some studies show no significant difference in cure rates between topical and oral antibiotics.

The recommended oral antibiotics include anti-staphylococcal penicillins, amoxicillin/clavulanate (Augmentin), cephalosporins, and macrolides. Erythromycin was found to be less effective.

Note that oral antibiotics can have more side effects than topical antibiotics, such as nausea.

Also, there is some evidence of antibiotic resistant staph in impetigo treatment.

Home Treatments

You can aid the healing and the appearance of impetigo with home treatments, cleaning and soaking and bleach baths.

Cleaning and soaking the sores is recommended, three to four times a day. Make sure to wash your hands thoroughly after treating the impetigo sores.

Gently clean the sores with warm water and soap and then remove the crusts from nonbullous impetigo. Removing the crusts exposes the bacteria underneath. You can also soak the affected area in warm soapy water before removing the crusts.

Cleaning or soaking and crust removal should be done regularly until the sores heal. Dry the area and apply antibiotic ointment. Then cover the sores lightly with gauze.

For a minor outbreak, you can use an over-the-counter antibiotic ointment.

Apply it three times a day, after cleaning the area. Then cover the sore with a bandage or gauze.

Another home treatment is a 15-minute bleach bath with a very dilute solution of household bleach (2.2 percent). This reduces the bacterial level on the skin, but needs to be done regularly.

For a full-size bath, use one-half cup of bleach. A full bath usually has 80 liters (21 gallons) of water. Rinse off with warm water and pat dry. Note that some people may have an allergic reaction to bleach.

A 2004 study found no evidence showing that other disinfecting agents, such as chlorhexidine or povidone-iodine, were effective. However, this study noted that more research was necessary.

Complications of Impetigo

Complications of impetigo can occur but are relatively rare. Generally, adults have a higher risk of complications.

About 1 to 5 percent of people with nonbullous impetigo get acute post-streptococcal glomerulonephritis, a serious disease involving inflammation of the small blood vessels in the kidneys.

Other complications of impetigo include:

- cellulitis, a serious infection (Staphlococcus aureus) of the tissues under your skin, which can spread to the bloodstream
- lymphangitis, an inflammation of the lymphatic channels
- sepsis, a bacterial infection of the blood
- scarlet fever, a rare bacterial infection caused by Streptococcus pyogenes
- guttate psoriasis, a non-infectious skin condition that can infect children and young adults after a skin infection
- Staphyloccus scalded skin syndrome (SSSS), another serious skin condition.

How can you Prevent Impetigo and its Spread

Children with impetigo should stay home until the impetigo is no longer contagious. Adults who have impetigo in the contagious stage and who work in occupations that involve close contact with others should check with their doctor about when to return to work.

Good hygiene is number one for prevention:

- Regular bathing and frequent handwashing can cut down on skin bacteria.
- Cover any skin wounds or insect bites to protect the area.
- Keep nails clipped and clean.
- Don't touch or scratch open sores. This will spread the infection.

- Wash everything that comes in contact with the impetigo sores in hot water and some laundry bleach.

- Change bed linens, towels, and clothing every day, until the sores are no longer contagious.

- Clean and disinfect surfaces, equipment, and toys that may have come in contact with impetigo.

- Don't share any personal items with someone who has impetigo.

Boils

A boil is a skin infection that starts in a hair follicle or oil gland. Also referred to as a skin abscess, it is a localized infection deep in the skin. A boil generally starts as a reddened, tender area. Over time, the area becomes firm and hard. Eventually, the center of the abscess softens and becomes filled with infection-fighting white blood cells that the body sends via the bloodstream to eradicate the infection. This collection of white blood cells, bacteria, and proteins is known as pus. Finally, the pus "forms a head," which can be surgically opened or spontaneously drain out through the surface of the skin.

Boil Symptoms

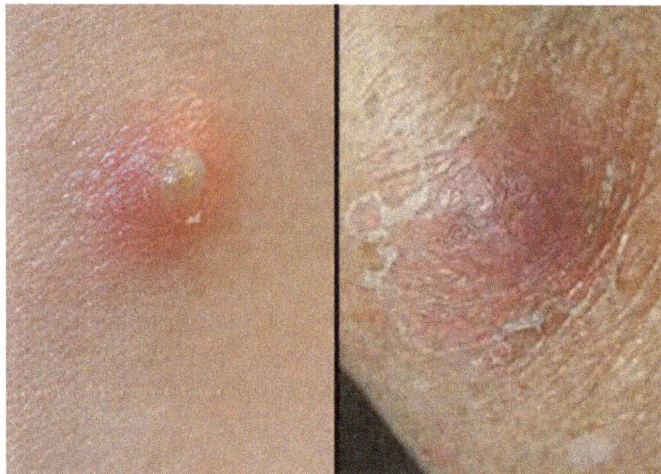

A boil starts as a hard, red, painful lump usually less than an inch in size. Over the next few days, the lump becomes softer, larger, and more painful. Soon a pocket of pus forms on the top of the boil. Signs of a severe infection are:

- The skin around the boil becomes red, painful, and swollen;

- more boils may appear around the original one;

- a fever develops;

- the lymph nodes in the area become swollen.

Places where Boils Form

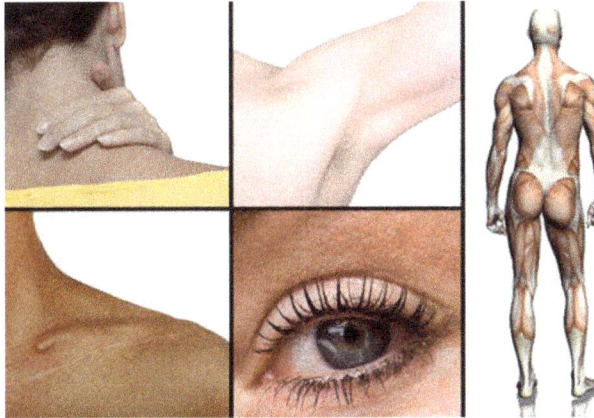

The most common places for boils to appear are on the:

- neck,

- armpits,

- shoulders,

- buttocks.

When one of these occur on the eyelid, it is called a sty (stye).

Causes of Boils

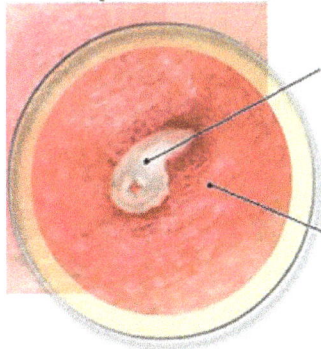

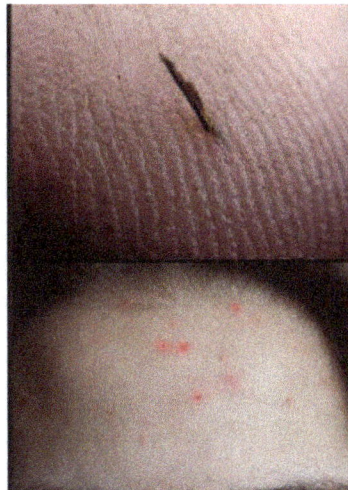

There are many causes of boils. Boils are usually caused by a type of bacteria called Staphylococcus (staph). Many staph infections develop into abscesses and can become serious very quickly. This germ can be present on normal skin and enters the body through tiny breaks in the skin or by traveling down a hair to the follicle. Some boils can be caused by an ingrown hair. Others can form as the result of a splinter or other foreign material that has become lodged in the skin that causes the infection to develop.

Additional Causes of Boils

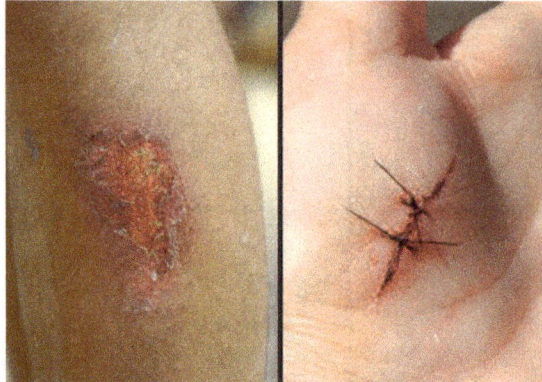

The skin is an essential part of our immune defense against materials and microbes that are foreign to our body. Any break in the skin, such as a cut or scrape, can develop into an abscess (boil) should it then become infected with bacteria.

Folliculitis could be an Early Warning

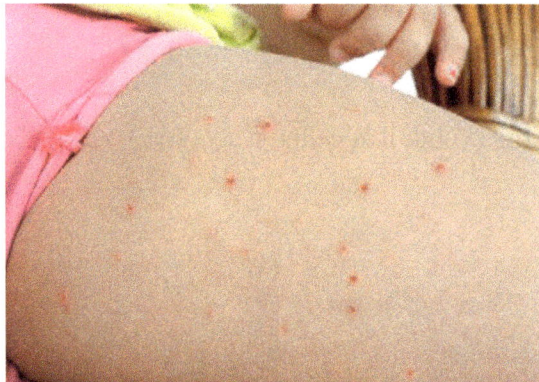

Folliculitis is an inflammation or infection of the hair follicles. This condition can develop into a boil and appears as numerous small red or pink little bumps at the hair follicles. Infection of the hair follicles can occur when the skin is disrupted or inflamed due to a number of conditions, including acne, skin wounds or injuries, friction from clothing, excessive sweating, or exposure to toxins.

Are Boils Contagious

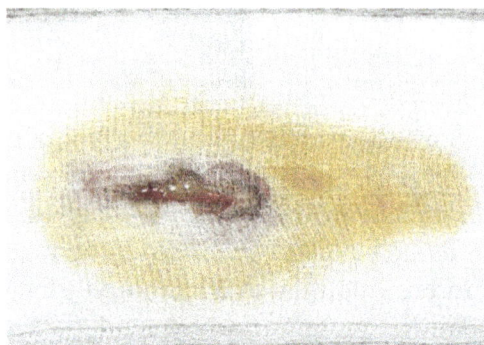

Boils themselves are not contagious, but the bacteria that cause boils are. Until it drains and heals, an active skin boil is contagious. The infection can spread to other parts of the person's body or to other people through skin-to-skin contact or the sharing of personal items.

Types of Boils

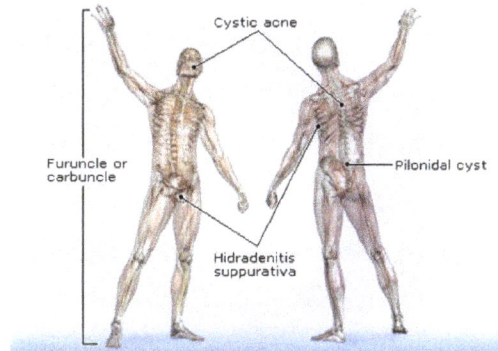

There are several different types of boils. Another name for a boil is furuncle. Among these are:

- carbuncle
- hidradenitis suppurativa (seen in the armpit or groin)
- pilonidal cyst
- cystic acne
- sty (stye)

Boil Type: Carbuncle

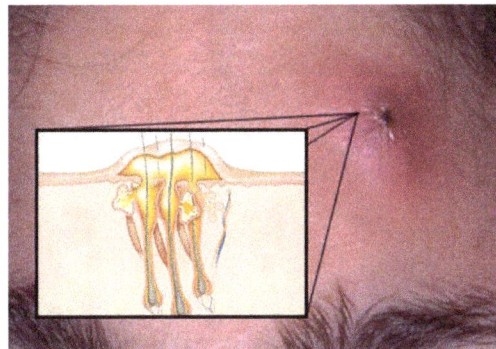

A carbuncle is an abscess in the skin caused by the bacterium Staphylococcus aureus. It usually involves a group of hair follicles and is therefore larger than a typical furuncle, or boil. A carbuncle can have one or more openings onto the skin and may be associated with fever or chills.

Boil Type: Cystic Acne

Cystic acne is a type of abscess that is formed when oil ducts become clogged and inflamed. Cystic acne affects deeper skin tissue than the more superficial inflammation from common acne. Cystic acne is most common on the face and typically occurs in the teenage years.

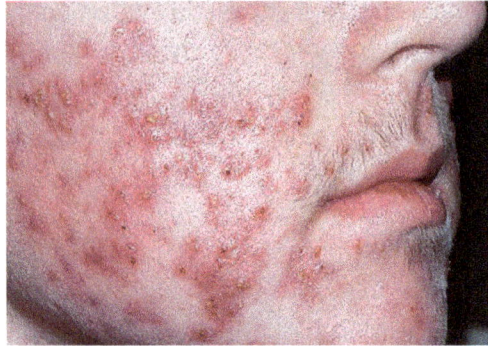

Boil Type: Hidradenitis Suppurativa

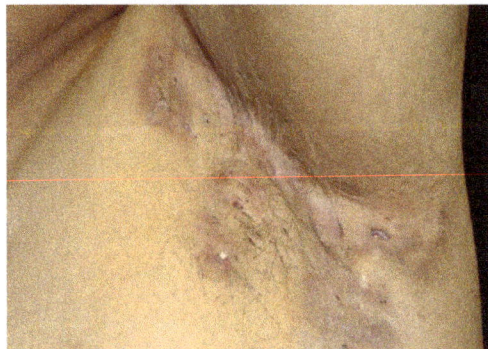

Hidradenitis suppurativa is a condition in which there are multiple abscesses that form under the armpits and often in the groin area. These areas are a result of local inflammation of the hair follicles. This form of skin inflammation is difficult to treat with antibiotics alone and typically requires a surgical procedure to remove the involved hair follicles in order to stop the skin inflammation.

Boil Type: Pilonidal Cyst

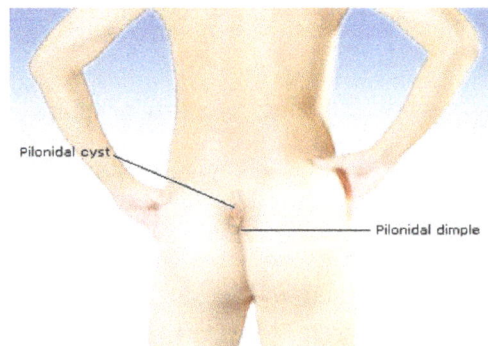

A pilonidal cyst is a unique kind of abscess that occurs in or above the crease of the buttocks. Pilonidal cysts often begin as tiny areas of inflammation in the base of the area of skin from which hair grows (the hair follicle). With irritation from direct pressure, over time the inflamed area enlarges to become a firm, painful, tender nodule making it difficult to sit without discomfort. These frequently form after long trips that involve prolonged sitting.

Boil Type: Sty

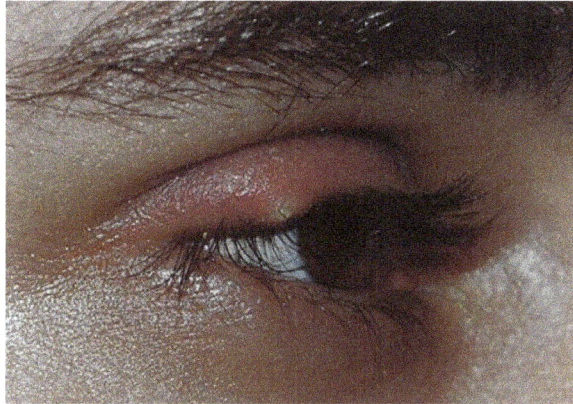

A sty (also spelled stye) is a tender, painful red bump located at the base of an eyelash or under or inside the eyelid. A sty results from a localized inflammation of the glands or a hair follicle of the eyelid. A sty is sometimes confused with a chalazion, a lump on the inner portion of the upper or lower eyelid, but a chalazion is usually painless and caused by obstruction and inflammation of an oil gland, not an infection.

People most likely to develop a Boil

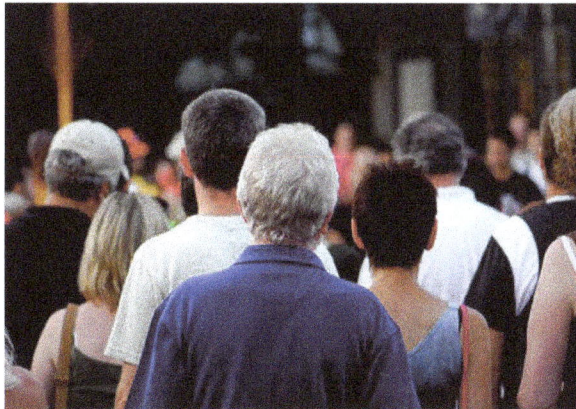

Anyone can develop a boil. However, people with certain illnesses or medications that impair the body's immune system are more likely to develop boils. Among the illnesses that can be associated with impaired immune systems are diabetes and kidney failure. Diseases, such as hypogamma-globulinemia, that are associated with deficiencies in the normal immune system, can increase the tendency to develop boils. Many medications can suppress the normal immune system and in-crease the risk of developing boils. These medications include cortisone medications (prednisone and prednisolone) and medications used for cancer chemotherapy.

Treatment for a Boil

Most simple boils can be treated at home. Ideally, the treatment should begin as soon as a boil is noticed since early treatment may prevent later complications. The primary treatment for most boils is heat application, usually with hot soaks or hot packs. Heat application increases the circulation to the area and allows the body to better fight off the infection by bringing antibodies and white blood cells to the site of infection. Do not pop the boil with a needle. This usually results in making the infection worse.

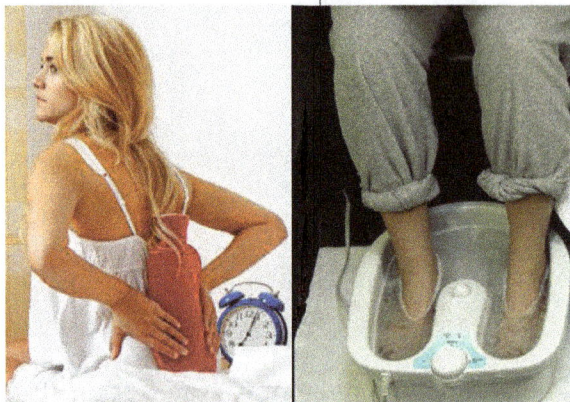

Should Boils be Drained

"="" it="" can="" be="" ready="" to="" drain."="">

As long as the boil is small and firm, opening the area and draining the boil is not helpful, even if the area is painful. However, once the boil becomes soft or "forms a head" (that is, a small pustule is noted in the boil), it can be ready to drain. Once drained, pain relief can be dramatic. Most small boils, such as those that form around hairs, drain on their own with soaking and/or heat application. On occasion, and especially with larger boils, the larger boil will need to be drained or "lanced" by a health-care practitioner. Frequently, these larger boils contain several pockets of pus that must be opened and drained.

Should Boils be Treated with Antibiotics

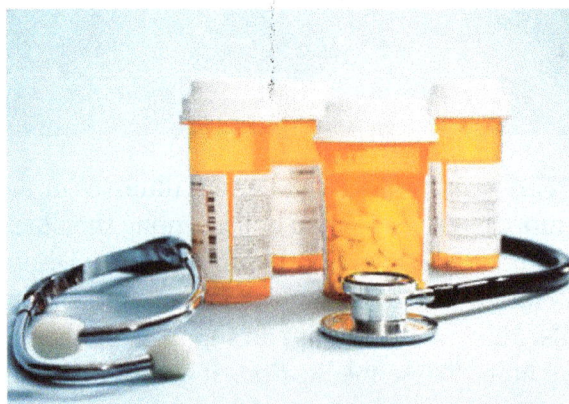

Antibiotics are often used to eliminate the accompanying bacterial infection. Especially if there is an infection of the surrounding skin, the doctor often prescribes antibiotics. However, antibiotics are not needed in every situation. In fact, antibiotics have difficulty penetrating the outer wall of an abscess and often will not cure an abscess without additional surgical drainage. In most cases, incising and draining the boil is sufficient to cure the infection.

Seeking Medical Attention

You should call your doctor and seek medical attention if:

- the boil is located on your face, near your spine, or near your anus;

- a boil is getting larger;

- the pain is severe;

- you have a fever;

- the skin around the boil turns red or red streaks appear;

- you have a heart murmur, diabetes, any problem with your immune system, or use immune-suppressing drugs (for example, corticosteroids or chemotherapy) and you develop a boil;

- the boil has not improved after five to seven days of home treatment;

- you get many boils over several months.

What can be done to Prevent Boils (Abscesses)

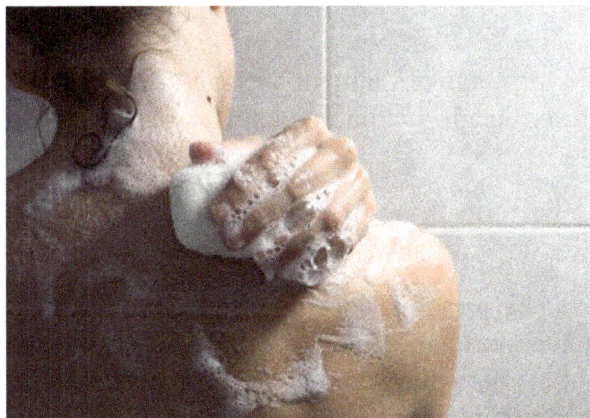

Good hygiene and the regular use of antibacterial soaps can help to prevent bacteria from building up on the skin. This can reduce the chance for hair follicles to become infected and prevent the formation of boils. Your health-care practitioner may recommend special cleansers such as Hibiclens to further reduce the bacteria on the skin.

Measures that can be taken to prevent more serious Boils and Abscesses

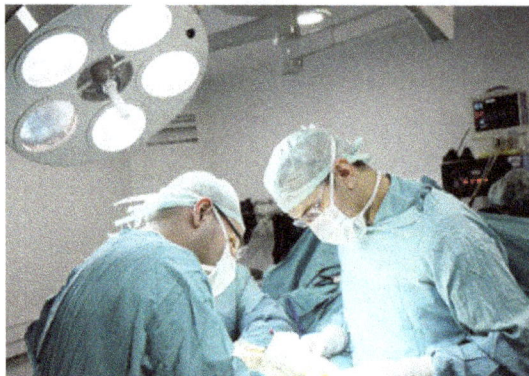

Pilonidal cysts can be prevented by avoiding continuous direct pressure or irritation of the buttock area when a local hair follicle becomes inflamed. Regular soap and hot water cleaning and drying can be helpful. For acne and hidradenitis suppurativa, antibiotics used as anti-inflammatory agents may be required on a long-term basis to prevent recurrent abscess formation. Finally, surgery may occasionally be needed, especially for hidradenitis suppurativa or pilonidal cysts that recur. For pilonidal cysts, surgically removing the outer shell of the cyst is important to clear the boil. For hidradenitis suppurativa, extensive involvement can require plastic surgery.

Campylobacter Jejuni

Campylobacteriosis is the disease caused by the infection with Campylobacter:

The onset of disease symptoms usually occurs 2 to 5 days after infection with the bacteria, but can range from 1 to 10 days.

The most common clinical symptoms of Campylobacter infections include diarrhoea (frequently bloody), abdominal pain, fever, headache, nausea, and/or vomiting. The symptoms typically last 3 to 6 days.

- Death from campylobacteriosis is rare and is usually confined to very young children or elderly patients, or to those already suffering from another serious disease such as AIDS.

- Complications such as bacteraemia (presence of bacteria in the blood), hepatitis, pancreatitis (infections of liver and pancreas, respectively), and miscarriage have been reported with various degrees of frequency. Post-infection complications may include reactive arthritis (painful inflammation of the joints which can last for several months) and neurological disorders such as Guillain-Barré syndrome, a polio-like form of paralysis that can result in respiratory and severe neurological dysfunction in a small number of cases.

Sources and Transmission

Campylobacter species are widely distributed in most warm-blooded animals. They are prevalent in food animals such as poultry, cattle, pigs, sheep and ostriches; and in pets, including cats and dogs. The bacteria have also been found in shellfish.

The main route of transmission is generally believed to be foodborne, via undercooked meat and meat products, as well as raw or contaminated milk. Contaminated water or ice is also a source of infection. A proportion of cases occur following contact with contaminated water during recreational activities.

Campylobacteriosis is a zoonosis, a disease transmitted to humans from animals or animal products. Most often, carcasses or meat are contaminated by Campylobacter from faeces during slaughtering. In animals, Campylobacter seldom causes disease.

The relative contribution of each of the above sources to the overall burden of disease is unclear but consumption of undercooked contaminated poultry is believed to be a major contributor. Since common-source outbreaks account for a rather small proportion of cases, the vast majority of reports refer to sporadic cases, with no easily discernible pattern.

Estimating the importance of all known sources is therefore extremely difficult. In addition, the wide occurrence of Campylobacter also hinders the development of control strategies throughout the food chain. However, in countries where specific strategies have been put in place to reduce the prevalence of Campylobacter in live poultry, a similar reduction in human cases is observed.

Treatment of C. jejuni Infections

Supportive measures, particularly fluid and electrolyte replacement are the principal therapies for most patients with campylobacteriosis. Severely dehydrated patients should receive rapid volume expansion with intravenous fluids. For most other patients, oral rehydration is indicated. Although Campylobacter infections are usually self limiting, antibiotic therapy may be prudent for patients who have high fever, bloody diarrhea, or more than eight stools in 24 hours; immunosuppressed patients, patients with bloodstream infections, and those whose symptoms worsen or persist for more than 1 week from the time of diagnosis. When indicated, antimicrobial therapy soon after the onset of symptoms can reduce the median duration of illness from approximately 10 days to 5 days. When treatment is delayed (e.g., until C. jejuni infection is confirmed by a medical laboratory), therapy may not be successful. Ease of administration, lack of serious toxicity, and high degree of efficacy make erythromycin the drug of choice for C. jejuni infection; however, other antimicrobial agents, particularly the quinolones and newer macrolides including azithromycin, are also used.

Antimicrobial Resistance

The increasing rate of human infections caused by antimicrobial-resistant strains of C. jejuni makes clinical management of cases of campylobacteriosis more difficult. Antimicrobial resistance can prolong illness and compromise treatment of patients with bacteremia. The rate of antimicrobial-resistant enteric infections is highest in the developing world, where the use of antimicrobial drugs in humans and animals is relatively unrestricted.

Pathogenesis

The pathogenesis of C. jejuni infection involves both host- and pathogen-specific factors. The health and age of the host and C. jejuni-specific humoral immunity from previous exposure influence clinical outcome after infection. In a volunteer study, C. jejuni infection occurred after ingestion of as few as 800 organisms. Rates of infection increased with the ingested dose. Rates of illness appeared to increase when inocula were ingested in a suspension buffered to reduce gastric acidity.

Many pathogen-specific virulence determinants may contribute to the pathogenesis of C. jejuni infection, but none has a proven role. Suspected determinants of pathogenicity include chemotaxis, motility, and flagella, which are required for attachment and colonization of the gut epithelium. Once colonization occurs, other possible virulence determinants are iron acquisition, host cell invasion, toxin production, inflammation and active secretion, and epithelial disruption with leakage of serosal fluid.

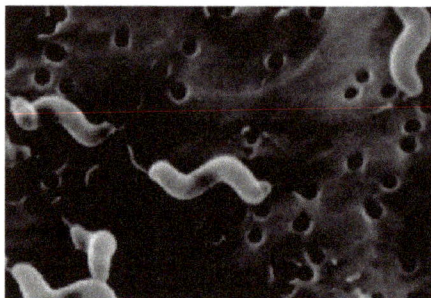

Figure: Scanning electron microscope image of Campylobacter jejuni, illustrating its corkscrew appearance and bipolar flagella.

Survival in the Environment

Survival of C. jejuni outside the gut is poor, and replication does not occur readily. C. jejuni grows best at 37°C to 42°C, the approximate body temperature of the chicken (41°C to 42°C). C. jejuni grows best in a low oxygen or microaerophilic environment, such as an atmosphere of 5% O_2, 10% CO_2, and 85% N_2. The organism is sensitive to freezing, drying, acidic conditions (pH < 5.0), and salinity.

Sample Collection and Transport

If possible, stool specimens should be chilled (not frozen) and submitted to a laboratory within 24 hours of collection. Storing specimens in deep, airtight containers minimizes exposure to oxygen and desiccation. If a specimen cannot be processed within 24 hours or is likely to contain small numbers of organisms, a rectal swab placed in a specimen transport medium (e.g., Cary-Blair) should be used. Individual laboratories can provide guidance on specimen handling procedures.

Numerous procedures are available for recovering C. jejuni from clinical specimens. Direct plating is cost-effective for testing large numbers of specimens; however, testing sensitivity may be reduced. Preenrichment (raising the temperature from 36°C to 42°C over several hours), filtration, or both are used in some laboratories to improve recovery of stressed C. jejuni organisms from specimens (e.g., stored foods or swabs exposed to oxygen). Isolation can be facilitated by using selective media containing antimicrobial agents, oxygen quenching agents, or a low oxygen atmosphere, thus decreasing the number of colonies that must be screened.

Subtyping of Isolates

No standard subtyping technique has been established for C. jejuni. Soon after the organism was described, two serologic methods were developed, the heat-stable or somatic O antigen and the heat-labile antigen schemes. These typing schemes are labor intensive, and their use is limited almost exclusively to reference laboratories. Many different DNA-based subtyping schemes have been developed, including pulsed-field gel electrophoresis (PFGE) and randomly amplified polymorphic DNA (RAPD) analysis. Various typing schemes have been developed on the basis of the sequence of flaA, encoding flagellin; however, recent evidence suggests that this locus may not be representative of the entire genome.

Transmission to Humans

Most cases of human campylobacteriosis are sporadic. Outbreaks have different epidemiologic characteristics from sporadic infections. Many outbreaks occur during the spring and autumn. Consumption of raw milk was implicated as the source of infection in 30 of the 80 outbreaks of human campylobacteriosis reported to CDC between 1973 and 1992. Outbreaks caused by drinking raw milk often involve farm visits (e.g., school field trips) during the temperate seasons. In contrast, sporadic Campylobacter isolates peak during the summer months. A series of case-control studies identified some risk factors for sporadic campylobacteriosis, particularly handling raw poultry and eating undercooked poultry. Other risk factors accounting for a smaller proportion of sporadic illnesses include drinking untreated water; traveling abroad; eating barbequed pork or sausage; drinking raw milk or milk from bird-pecked bottles; and contact with dogs and cats, particularly juvenile pets or pets with diarrhea.

Reservoirs

The ecology of C. jejuni involves wildlife reservoirs, particularly wild birds. Species that carry C. jejuni include migratory birds—ranes, ducks, geese, and seagulls. The organism is also found in other wild and domestic bird species, as well as in rodents. Insects can carry the organism on their exoskeleton.

The intestines of poultry are easily colonized with C. jejuni. Day-old chicks can be colonized with as few as 35 organisms. Most chickens in commercial operations are colonized by 4 weeks. Vertical transmission (i.e., from breeder flocks to progeny) has been suggested in one study but is not widely accepted. Reservoirs in the poultry environment include beetles, unchlorinated drinking water, and farm workers. Feeds are an unlikely source of campylobacters since they are dry and campylobacters are sensitive to drying.

C. jejuni is a commensal organism of the intestinal tract of cattle. Young animals are more often colonized than older animals, and feedlot cattle are more likely than grazing animals to carry campylobacters. In one study, colonization of dairy herds was associated with drinking unchlorinated water.

Campylobacters are found in natural water sources throughout the year. The presence of campylobacters is not clearly correlated with indicator organisms for fecal contamination (e.g., E. coli). In temperate regions, organism recovery rates are highest during the cold season. Survival in cold water is important in the life cycle of campylobacters. In one study, serotypes found in water were similar to those found in humans. When stressed, campylobacters enter a "viable but nonculturable state," characterized by uptake of amino acids and maintenance of an intact outer membrane but inability to grow on selective media; such organisms, however, can be transmitted to animals. Additionally, unchlorinated drinking water can introduce campylobacters into the farm environment.

Campylobacter in the Food Supply

C. jejuni is found in many foods of animal origin. Surveys of raw agricultural products support epidemiologic evidence implicating poultry, meat, and raw milk as sources of human infection. Most retail chicken is contaminated with C. jejuni; one study reported an isolation rate of 98% for retail chicken meat. C. jejuni counts often exceed 103 per 100 g. Skin and giblets have particularly high levels of contamination. In one study, 12% of raw milk samples from dairy farms in eastern Tennessee were contaminated with C. jejuni. Raw milk is presumed to be contaminated by bovine feces; however, direct contamination of milk as a consequence of mastitis also occurs. Campylobacters are also found in red meat. In one study, C. jejuni was present in 5% of raw ground beef and in 40% of veal specimens.

Control of Campylobacter Infection

On the Farm

Control of Campylobacter contamination on the farm may reduce contamination of carcasses, poultry, and red meat products at the retail level. Epidemiologic studies indicate that strict hygiene reduces intestinal carriage in food-producing animals. In field studies, poultry flocks that drank chlorinated water had lower intestinal colonization rates than poultry that drank unchlorinated water. Experimentally, treatment of chicks with commensal bacteria and immunization of older birds reduced C. jejuni colonization. Because intestinal colonization with campylobacters readily occurs in poultry flocks, even strict measures may not eliminate intestinal carriage by food-producing animals.

At Processing

Slaughter and processing provide opportunities for reducing C. jejuni counts on food-animal carcasses. Bacterial counts on carcasses can increase during slaughter and processing steps. In one study, up to a 1,000-fold increase in bacterial counts on carcasses was reported during transportation to slaughter. In studies of chickens and turkeys at slaughter, bacterial counts increased by approximately 10- to 100-fold during defeathering and reached the highest level after evisceration. However, bacterial counts on carcasses decline during other slaughter and processing steps. In one study, forced-air chilling of swine carcasses caused a 100-fold reduction in carcass contamination. In Texas turkey plants, scalding reduced carcass counts to near or below detectable levels. Adding sodium chloride or trisodium phosphate to the chiller water in the presence of an electrical current reduced C. jejuni contamination of chiller water by 2 log10 units. In a slaughter plant in England, use of chlorinated sprays and maintenance of clean working surfaces resulted in a 10- to 100-fold decrease in carcass contamination. In another study, lactic acid spraying of swine carcasses reduced counts by at least 50% to often undetectable levels. A radiation dose of 2.5 KGy reduced C. jejuni levels on retail poultry by 10 log10 units.

Botulinum

Botulism is a rare but serious condition caused by toxins from bacteria called Clostridium botulinum.

Three common forms of botulism are:

- Foodborne botulism: The harmful bacteria thrive and produce the toxin in environments with little oxygen, such as in home-canned food.

- Wound botulism: If these bacteria get into a cut, they can cause a dangerous infection that produces the toxin.

- Infant botulism: This most common form of botulism begins after Clostridium botulinum bacterial spores grow in a baby's intestinal tract. It typically occurs in babies between the ages of 2 months and 8 months.

All types of botulism can be fatal and are considered medical emergencies.

Symptoms

Foodborne Botulism

Signs and symptoms of foodborne botulism typically begin between 12 and 36 hours after the toxin gets into your body. But, depending on how much toxin was consumed, the start of symptoms may range from a few hours to a few days. Signs and symptoms of foodborne botulism include:

- Difficulty swallowing or speaking

- Dry mouth

- Facial weakness on both sides of the face

- Blurred or double vision

- Drooping eyelids

- Trouble breathing

- Nausea, vomiting and abdominal cramps

- Paralysis

Wound Botulism

Signs and symptoms of wound botulism appear about 10 days after the toxin has entered the body. Wound botulism signs and symptoms include:

- Difficulty swallowing or speaking

- Facial weakness on both sides of the face

- Blurred or double vision

- Drooping eyelids
- Trouble breathing
- Paralysis

The wound may or may not appear red and swollen.

Infant Botulism

If infant botulism is related to food, such as honey, problems generally begin within 18 to 36 hours after the toxin enters the baby's body. Signs and symptoms include:

- Constipation, which is often the first sign
- Floppy movements due to muscle weakness and trouble controlling the head
- Weak cry
- Irritability
- Drooling
- Drooping eyelids
- Tiredness
- Difficulty sucking or feeding
- Paralysis

Certain signs and symptoms usually don't occur with botulism. For example, botulism doesn't generally increase blood pressure or heart rate, or cause fever or confusion. Sometimes, however, wound botulism may cause fever.

When to See a Doctor

Seek urgent medical care if you suspect that you have botulism. Early treatment increases your chances of survival and lessens your risk of complications.

Seeking medical care promptly may also alert public health authorities. They may then be able to keep other people from eating contaminated food. Botulism isn't contagious from person to person.

Causes

Foodborne Botulism

The source of foodborne botulism is often home-canned foods that are low in acid, such as fruits, vegetables and fish. However, the disease has also occurred from spicy peppers (chiles), foil-wrapped baked potatoes and oil infused with garlic.

When you eat food containing the toxin, it disrupts nerve function, causing paralysis.

Wound Botulism

When C. botulinum bacteria get into a wound possibly caused by an injury you might not notice they can multiply and produce toxin. Wound botulism has increased in recent decades in people who inject heroin, which can contain spores of the bacteria. In fact, this type of botulism is more common in people who inject black tar heroin.

Infant Botulism

Babies get infant botulism after consuming spores of the bacteria, which then grow and multiply in their intestinal tracts and make toxins. The source of infant botulism may be honey, but it's more likely to be exposure to soil contaminated with the bacteria.

Benefits of Botulinum Toxin

You might wonder how something so toxic could ever be beneficial, but scientists have found that the paralyzing effect of botulinum toxin makes it useful in certain circumstances.

Botulinum toxin has been used to reduce facial wrinkles by preventing contraction of muscles beneath the skin and for medical conditions, such as eyelid spasms and severe headaches. However, there have been rare occurrences of serious side effects, such as muscle paralysis extending beyond the treated area, with the use of botulinum toxin for medical reasons. Be sure to use a licensed doctor for any cosmetic or medical procedures using onabotulinumtox-inA (Botox).

Complications

Because it affects muscle control throughout your body, botulinum toxin can cause many complications. The most immediate danger is that you won't be able to breathe, which is the most common cause of death in botulism. Other complications, which may require rehabilitation, may include:

- Difficulty speaking
- Trouble swallowing
- Long-lasting weakness
- Shortness of breath

Prevention

Use Proper Canning Techniques

Be sure to use proper techniques when canning foods at home to ensure that any botulism germs in the food are destroyed:

- Pressure-cook these foods at 250 F (121 C) for 20 to 100 minutes, depending on the food.
- Consider boiling these foods for 10 minutes before serving them.

Prepare and Store Food Safely

- Don't eat preserved food if its container is bulging or if the food smells spoiled. However, taste and smell won't always give away the presence of C. botulinum. Some strains don't make food smell bad or taste unusual.

- If you wrap potatoes in foil before baking them, eat them hot or loosen the foil and store them in the refrigerator not at room temperature.

- Store oils infused with garlic or herbs in the refrigerator.

Infant Botulism

To reduce the risk of infant botulism, avoid giving honey even a tiny taste to children under the age of 1 year.

Wound Botulism

To prevent wound botulism and other serious bloodborne diseases, never inject or inhale street drugs.

Diagnosis

To diagnose botulism, your doctor will check you for signs of muscle weakness or paralysis, such as drooping eyelids and a weak voice. Your doctor will also ask about the foods you've eaten in the past few days, and ask if you may have been exposed to the bacteria through a wound.

In cases of possible infant botulism, the doctor may ask if the child has eaten honey recently and has had constipation or sluggishness.

Analysis of blood, stool or vomit for evidence of the toxin may help confirm an infant or foodborne botulism diagnosis. But because these tests may take days, your doctor's exam is the main way to diagnose botulism.

Treatment

For cases of foodborne botulism, doctors sometimes clear out the digestive system by inducing vomiting and giving medications to induce bowel movements. If you have botulism in a wound, a doctor may need to remove infected tissue surgically.

Antitoxin

If you're diagnosed early with foodborne or wound botulism, injected antitoxin reduces the risk of complications. The antitoxin attaches itself to toxin that's still circulating in your bloodstream and keeps it from harming your nerves.

The antitoxin cannot, however, reverse the damage that's been done. Fortunately, nerves do re-generate. Many people recover fully, but it may take months and extended rehabilitation therapy.

A different type of antitoxin, known as botulism immune globulin, is used to treat infants.

Antibiotics

Antibiotics are recommended for the treatment of wound botulism. However, these medications are not advised for other types of botulism because they can speed up the release of toxins.

Breathing Assistance

If you're having trouble breathing, you'll probably need a mechanical ventilator for as long as several weeks as the effects of the toxin gradually lessen. The ventilator forces air into your lungs through a tube inserted in your airway through your nose or mouth.

Rehabilitation

As you recover, you may also need therapy to improve your speech, swallowing and other functions affected by the disease.

Listeria Monocytogenes

Listeria monocytogenes is a bacterium that causes listeriosis, a disease that can have severe consequences for particular groups of the population. It can cause miscarriages in pregnant women and be fatal in immunocompromised individuals and the elderly. In healthy people, listeriosis generally only causes a mild form of illness. L. monocytogenes can be found throughout the environment. It has been isolated from domestic and wild animals, birds, soil, vegetation, fodder, water and from floors, drains and wet areas of food processing factories.

Description of the Organism

L. monocytogenes is a Gram-positive, non-spore forming rod-shaped bacterium. It belongs to the genus Listeria along with L. ivanovii, L. innocua, L. welshimeri, L. selligeri and L. grayi Rocourt and Buchrieser 2007. Of these species, only two are considered pathogens: L. monocytogenes which infects humans and animals, and L. ivanovii which infects ruminants (although there have been rare reports of L. ivanovii being isolated from infected humans) Guillet et al. 2010. There are thirteen known serotypes of L. monocytogenes: 1/2a, 1/2b, 1/2c, 3a, 3b, 3c, 4a, 4ab, 4b, 4c, 4d, 4e and 7. The serotypes most often associated with human illness are 1/2a, 1/2b and 4b (FDA 2012).

Growth and Survival Characteristics

The growth and survival of L. monocytogenes is influenced by a variety of factors. In food these include temperature, pH, water activity, salt and the presence of preservatives.

The temperature range for growth of L. monocytogenes is between -1.5 and 45°C, with the optimal temperature being 30–37°C. Temperatures above 50°C are lethal to L.monocytogenes. Freezing can also lead to a reduction in L. monocytogenes numbers. As L. monocytogenes can grow at temperatures as low as 0°C, it has the potential to grow, albeit slowly, in food during refrigerated storage.

L. monocytogenes will grow in a broad pH range of 4.0–9.6. Although growth at pH <4.0 has not been documented, L. monocytogenes appears to be relatively tolerant to acidic conditions. L. monocytogenes becomes more sensitive to acidic conditions at higher temperatures.

Like most bacterial species, L. monocytogenes grows optimally at a water activity (aw) of 0.97. However, L. monocytogenes also has the ability to grow at a a_w of 0.90. Johnson demonstrated that L. monocytogenes can survive for extended periods of time at a a_w value of 0.81. L. monocytogenes is reasonably tolerant to salt and has been reported to grow in 13–14% sodium chloride. Survival in the presence of salt is influenced by the storage temperature. Studies have indicated that in concentrated salt solutions, the survival rate of L. monocytogenes is higher when the temperature is lower.

L. monocytogenes can grow under both aerobic and anaerobic conditions, although it grows better in an anaerobic environment.

The effect of preservatives on the growth of L. monocytogenes is influenced by the combined effects of temperature, pH, salt content and water activity. For example, sorbates and parabens are more effective at preventing growth of L. monocytogenes at lower storage temperatures and pH. Also, adding sodium chloride or lowering the temperature enhances the ability of lactate to prevent L. monocytogenes growth. At decreased temperatures (such as refrigeration storage) sodium diacetate, sodium propionate and sodium benzoate are more effective at preventing growth of L. monocytogenes.

Table: Limits for growth of L. monocytogenes when other conditions are near optimum.

	Minimum	Optimum	Maximum
Temperature (°C)	-1.5	30–37	45
pH	4.0	6.0–8.0	9.6
Water activity	0.90	0.97	-

Symptoms of Disease

There are two main forms of illness associated with L. monocytogenes infection. Noninvasive listeriosis is the mild form of disease, while invasive listeriosis is the severe form of disease and can be fatal. The likelihood that invasive listeriosis will develop depends upon a number of factors, including host susceptibility, the number of organisms consumed and the virulence of the particular strain.

Symptoms of non-invasive listeriosis can include fever, diarrhoea, muscle aches, nausea, vomiting, drowsiness and fatigue. The incubation period is usually 1 day (range 6 hours to 10 days). Non-invasive listeriosis is also known as listerial gastroenteritis or febrile listeriosis.

Invasive listeriosis is characterised by the presence of L. monocytogenes in the blood, in the fluid of the central nervous system (leading to bacterial meningitis) or infection of the uterus of pregnant women. The latter may result in spontaneous abortion or stillbirth (20% of cases) or neonatal infection (63% of cases). Influenza-like symptoms, fever and gastrointestinal symptoms often occur in pregnant women with invasive listeriosis. In non-pregnant adults, invasive listeriosis presents in the form of bacterial meningitis with a fatality rate of 30%. Symptoms including fever, malaise, ataxia, seizures and altered mental status. The incubation period before onset of invasive listeriosis ranges from 3 days to 3 months.

Virulence and Infectivity

When L. monocytogenes is ingested, it may survive the stomach environment and enter the intestine where it penetrates the intestinal epithelial cells. The organism is then taken up by macrophages and non-phagocytic cells. The L. monocytogenes surface protein internalin is required for this uptake by non-phagocytic cells, as it binds to the receptors on the host cells to instigate adhesion and internalization. The bacterium is initially located in a vacuole after uptake by a macrophage or non-phagocytic cell. L. monocytogenes secrete listeriolysin O protein, which breaks down the vacuole wall and enables the bacteria to escape into the cytoplasm. Any bacteria remaining in the vacuole are destroyed by the host cell. Once located in the cytoplasm of the host cell, L. monocytogenes is able to replicate. L. monocytogenes is transported around the body by the blood, with most L. monocytogenes being inactivated when it reaches the spleen or liver. L. monocytogenes is able to utilise the actin molecules of the host to propel the bacteria into neighbouring host cells. In the case of invasive listeriosis, this ability to spread between host cells enables L. monocytogenes to cross the blood-brain and placental barriers.

Mode of Transmission

The most common transmission route of L. monocytogenes to humans is via the consumption of contaminated food. However, L. monocytogenes can be transmitted directly from mother to child (vertical transmission), from contact with animals and through hospital acquired infections.

Healthy individuals can be asymptomatic carriers of L. onocytogenes, with 0.6–3.4% of healthy people with unknown exposure to Listeria being found to shed L. monocytogenes in their faeces. However, outbreak investigations have shown that listeriosis patients do not always shed the organism in their faeces. Therefore the role of healthy carriers in the transmission of L. monocytogenes is unclear.

Occurrence in Food

The presence of L. monocytogenes in ready-to-eat products is probably due to contamination occurring after the product has been processed. This contamination may occur during additional handling steps such as peeling, slicing and repackaging. Also, in the retail and food service environment, contamination may be transferred between ready-to-eat products Lianou and Sofos 2007. The type of handling that ready-to-eat meat receives may also influence the level of L. monocytogenes contamination. In a survey of retail packaged meats there was a significantly higher prevalence of L. monocytogenes reported in products cut into cubes (61.5%) (n=13), compared with sliced products (4.6%) (n=196).

Host Factors that Influence Disease

People at risk of invasive listeriosis include pregnant women and their foetuses, newborn babies, the elderly and immune compromised individuals (such as cancer, transplant and HIV/AIDS patients). Less frequently reported, but also at a greater risk, are patients with diabetes, asthma, cirrhosis (liver disease) and ulcerative colitis (inflammatory bowel disease).

Dose Response

Investigations of foodborne outbreaks of non-invasive listeriosis have concluded that consumption of food with high levels of L. monocytogenes (1.9×10^5 /g to 1.2×10^9 /g) is required to cause illness in the general healthy population.

The number of L. monocytogenes required to cause invasive listeriosis depends on a number of factors. These include the virulence of the particular serotype of L. monocytogenes, the general health and immune status of the host, and attributes of the food (for example fatty foods can protect bacteria from stomach acid). Some L. monocytogenes serovars are more virulent than others; this may be attributed to differences in the expression of virulence factors which could influence the interactions between the bacterium and the host cells and cellular invasion. The FDA and WHO have developed separate models for both healthy and susceptible populations to predict the probability that an individual will develop listeriosis. The probability that a healthy person of intermediate age will become ill from the consumption of a single L. monocytogenes cell was estimated to be 2.37×10^{-14}. For more susceptible populations the probability that illness will occur was estimated to be 1.06×10^{-12}. A more recent assessment on invasive listeriosis in susceptible populations was performed which took into account the different serotypes of L. monocytogenes Chen et al. 2006. This study showed that the probability of a susceptible individual developing invasive listeriosis ranged from 1.31×10^{-8} to 5.01×10^{-11}, suggesting that there are large differences in virulence between L. monocytogenes serotypes.

Salmonella

- Salmonella is a group of bacteria that normally inhabit the intestines of animals and humans. Almost all warm- and cold-blooded animals, including dogs,cats, rabbits, rodents and other small pocket pets, reptiles, birds and livestock (e.g. cattle, horses, poultry, swine) can carry or be infected by Salmonella of some kind. Disease due to Salmonella infection is called salmonellosis.

- The most important types (also called serotypes) of Salmonella in human and veterinary medicine mostly belong to the same subspecies, Salmonella enterica subsp. enterica. Therefore, unlike other bacteria, the species and subspecies are often not written, and the strains are simply refered to by their serotype, such as Salmonella Enteriditis, S. Heidelberg and S. Typhimurium.

  o Salmonella Typhi, the serotype that causes typhoid fever in humans, does not infect animals and is therefore not a zoonotic pathogen.

- Salmonella is an important cause of disease in humans and pets. The most common sign of illness is diarrhea. However, Salmonella can also be carried by humans and pets without any signs of illness at all.

- Salmonella can sometimes spread beyond the intestinal tract, resulting in severe, even life-threatening infection of other parts of the body, particularly in animals or people who are very young, old, or have a weakened immune system.

- Salmonella is so commonly found in reptiles that all reptiles should be assumed to be carrying Salmonella, and handled appropriately.

- The risk of transmission of Salmonella between animals and people can be reduced by increasing awareness of how it is transmitted and some common-sense infection control measures.

How Common is Salmonella

Humans

Salmonellosis is one of the most common causes of bacterial diarrhea in humans worldwide. In Canada, approximately 6,000-12,000 cases of salmonellosis are reported yearly, but because diarrheal diseases are typically under-reported, the true number of cases may be closer to hundreds of thousands.

Most cases of salmonellosis are attributed to food poisoning, but outbreaks have also been linked to exposure to contaminated pet food and pet treats, exposure to livestock, and exposure to pets, particularly reptiles.

Salmonellosis can affect anyone, including healthy adults. However, the young, elderly or immunocompromised individuals (e.g. HIV/AIDS, cancer or transplant patients) are at higher risk of developing disease and serious complications when exposed to Salmonella.

Animals

- Recent studies estimate the between 1% to 4% of healthy dogs, and less than 1% up to 18% in healthy cats may carry Salmonella in their intestine. The majority of dogs and cats with Salmonella show no signs of infection.

- Salmonella in dogs and cats appears to be more common in animals that are housed closely together in high numbers such as kennels, catteries, and shelters.

- Risk factors for Salmonella carriage in dogs and cats include raw meat diets and exposure to livestock.

- Salmonella in pocket pets such as mice, rats, gerbils, hamsters, and guinea pigs is likely also quite common.In 2004, an outbreak of human salmonellosis in the USA was linked to hamsters. Salmonella is very commonly found in reptiles, and reptile exposure has caused multiple outbreaks of salmonellosis in humans.

- There is also evidence that Salmonella can be transmitted from dogs and cats to humans in some instances.

- Adult dogs and cats seem to be somewhat naturally resistant to disease due to Salmonella. Risk factors for clinical salmonellosis that have been identified during hospital outbreaks in dogs include pre-existing disease, prolonged hospitalization, major surgery, steroid therapy, cancer, chemotherapy, and antibiotic therapy.

The way in which Salmonella infects Animals and People

Salmonella normally lives in the intestine of humans and animals. However, Salmonella is a very hardy organism, and can potentially be found almost anywhere in soil, water, or anywhere contaminated with animal stool or human sewage and it can survive for long periods of time, particularly in the presence of organic matter.

- Salmonella is usually transmitted by swallowing contaminated water or food, or contamination of the hands with stool which is then transferred to the mouth.

  o Contamination of food is not limited to products of animal origin. For example, fruits and vegetables which have been irrigated or fertilized with contaminated water or manure, and which are not washed with potable water or cooked prior to being eaten, can also be an important source of Salmonella.

- Foodborne transmission (i.e. "food-poisoning") is a much more common cause of illness due to salmonellosis, but contact with stool from either wild or domestic animals can also result in this disease.

- Animals may be infected in the same way as people – by consuming contaminated water or food, or swallowing Salmonella after licking or chewing a contaminated object.

- The number of Salmonella organisms that need to be swallowed in order to cause illness depends on the strain of Salmonella involved and the health status and age of the animal or person.

Effects of Salmonellosis on a Person or Animal

Humans: Signs of salmonellosis in people can range from none to severe, sudden diarrhea. In some cases, the infection can also spread to tissues in the body beyond the intestine, resulting in complications such as bloodstream infection, abscesses, meningitis, and infection of bone, joints and heart valves. People with weakened immune systems (e.g. HIV/AIDS, cancer or transplant patients), young children or elderly individuals are at a particularly high risk for severe disease and complications associated with Salmonella infection, and in some cases the infection can be fatal, especially in infants. In most individuals with uncomplicated diarrhea, the infection resolves on its own, but chronic infection can occur in some cases, resulting in similar but intermittent signs over a period of months. Affected individuals, both human and animal, typically shed Salmonella in their stool even after signs of illness resolve, sometimes for weeks. Animals: Salmonella infection does not usually cause visible signs of illness in pets, but it can cause acute, intermittent or chronic diarrhea, just like it does in people. Infection can also spread beyond the intestinal tract, causing infection of the bloodstream and other organs. Young animals, stressed/sick animals or those with a weakened immune system, and those kept in large groups such as kennels or shelters are at higher risk for getting sick. Over 90% of dogs and cats recover from acute salmonellosis.

Diagnosis of Salmonellosis

In both animals and people, the diagnosis of salmonellosis is typically made by culturing the bacteria from the stool. However, because Salmonella can be found in the intestines of healthy individuals, a positive culture alone does not necessarily mean that Salmonella is the cause of the person's or animal's disease. Isolation of Salmonella from tissues or fluid not associated with the intestinal tract (e.g. blood, lung secretions, urine) is more definitive, and indicates that the infection has invaded the rest of the body.

Shedding of Salmonella is intermittent in animals and people. The number of bacteria in the stool and how often they are passed varies from person to person and animal to animal. Therefore testing of multiple fecal samples may be necessary for diagnosis.

Treatment of Salmonellosis

- Diarrhea: Straightforward cases of diarrhea due to salmonellosis usually do NOT require antibiotic treatment – the infection typically resolves on its own, although oral or intravenous fluids may be needed to prevent dehydration if the diarrhea is severe. Antibiotic therapy in salmonellosis is only meant to prevent spread of bacteria to the rest of the body, not to kill Salmonella within the intestinal tract. Since spread of the infection within the body is uncommon in otherwise healthy people or animals, antibiotics are likely unnecessary, and may actually lead to increased antimicrobial resistance, and may prolong the shedding period after the patient recovers.

- Treatment of any kind must always be accompanied by infection control measures to prevent the spread of Salmonella and prevent re-infection of the person or animal. It is not possible to predict how long a person or animal will shed Salmonella following infection, but shedding may persist intermittently for weeks. The duration of shedding is likely influenced by the patients overall health and the condition of the normal bacterial population in the intestine.

Systemic infection: In animals and humans with infection of the bloodstream of other tissues, antibiotics are needed, as well as other treatments depending on organ systems are affected. Systemic infections are usually very serious, and may require hospitalization.

Healthy pets that are shedding Salmonella should not be treated with antibiotics. There is no evidence that antibiotic treatment helps pets stop shedding Salmonella sooner, and it may actually prolong the amount of time an animal sheds the bacteria, because antibiotics disrupt the normal bacterial population in the intestine. Unnecessary use of antibiotics can also lead to antibiotic resistance. Probiotics have also not been shown to be effective for eliminating Salmonella shedding in animals.

Salmonella vaccines are not available for dogs, cats, or humans. An effective vaccine for dogs and cats may not be possible due to the large number of different serotypes of Salmonella that may infect these species.

Preventing Pets from getting Salmonella

Completely preventing pets from carrying Salmonella can be very frustrating, and is likely impossible, because animals can carry the bacteria without signs of illness, and the bacteria survive so well in the environment. However, taking steps to control the transmission of Salmonella in pets and people is still very important from a human and public health perspective. Any Salmonella infection of an animal, whether the animal is sick or not, should be considered potentially transmissible to humans, and vice versa. Control of stool contamination, both human and animal, is the most important preventative measure.

Hand Hygiene: Anyone handling a pet or stool from a pet should wash their hands immediately afterwards with soap and running water, or use an alcohol-based hand sanitizer. This is especially important in the case of animals with confirmed or suspected Salmonella infection (e.g. a pet with diarrhea, any pet that is fed raw meat), but applies to all animals, as even healthy pets can shed Salmonella. Hand hygiene is also critical:

- After using the bathroom
- Prior to handling any food
- After handling any kind of raw meat product
- After handling any kind of pet food.

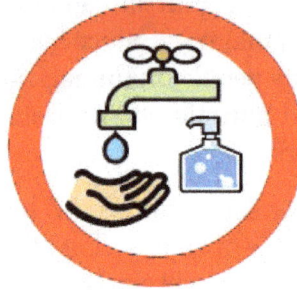

At Ome and in Public

Dog stool should be picked up immediately to prevent environmental contamination, especially in public areas like parks where other dogs and children may play.

- Prevent pets from drinking from puddles, ponds, lakes or other water sources that may be contaminated with feces from other animals.

- Dogs should be strongly discouraged from eating their own stool or that of other animals.

- Thoroughly cook all food (especially meat) fed to pets.

- Feeding a commercially prepared, heat-processed diet helps to reduce the risk of Salmonella contamination in the food, but even these products can occasionally contain Salmonella. Pet food should therefore be kept in a sealed container and should never come in contact with kitchen surfaces or food meant for human consumption.

- Do not leave wet pet food in dishes at room temperature for prolonged periods, as this provides ideal conditions for bacteria of many kinds to grow.

- Prevent pets from hunting and scavenging small wild animals and birds. Dogs should be supervised carefully when off-leash. Cats should ideally be kept indoors.

Measures that can be taken when pets are diagnosed with Salmonellosis

High-risk individuals (e.g. children, the elderly or persons with a weakened immune system) should avoid contact with diarrheic pets, as these animals may shed high numbers of potentially zoonotic pathogens, including but not limited to Salmonella, in their stool. Members of households that include high-risk individuals must pay particularly close attention to hand hygiene and other infection control measures at all times, even if the pet in question is healthy, as any animal (or person) could potentially be shedding Salmonella in its stool.

- Preventing stool contamination of the environment, the pet's haircoat, and the hands and clothing of people in contact with the animal is of primary importance. Diligent attention to hand hygiene, cleaning and disinfection of any surface that becomes contaminated with pet feces are crucial. Linens that become contaminated should be washed separately and dried completely using high heat in a dryer.

- Most disinfectants, including a simple 1:10 solution of household bleach, can effectively kill Salmonella if all visible organic debris is removed beforehand, and adequate contact time (10-15 minutes) is allowed.

Animals Fed Raw Meat

Feeding raw meat to pets significantly increases the risk of Salmonella carriage. Raw meat diets and stool from animals fed these diets may pose a risk to other animals and people. Pets that live or have frequent contact with high risk individuals (e.g. very young or elderly persons, or those with weakened immune systems) should NOT be fed raw meat.

Pet treats made from raw animal tissues may also be contaminated with Salmonella, and outbreaks associated with these products have been reported. Raw treats such as pig ears and rawhides should be handled in the same manner as raw meat, and should not be given to dogs in households with high-risk individuals.

Therapy Animals

Pets that visit healthcare facilities or are part of other animal visitation programs should not be fed raw meat due to the increased risk of shedding Salmonella. Guidelines have been developed to reduce the risk of pets involved in animal visitation programs acquiring or transmitting infectious diseases. If your pet is involved in these programs, ensure that you follow these guidelines.

An examination of whether a pet needs to be tested if a person is diagnosed with Salmonellosis

People can potentially transmit Salmonella to pets and other people. Anyone diagnosed with salmonellosis should be very diligent about washing their hands thoroughly after using the bathroom, and pet(s) should be prevented from drinking from the toilet. There is no evidence that testing pets for Salmonella is useful if a person in the household is diagnosed with salmonellosis.

Reptiles & Amphibians

These animals are of considerably higher risk for transmission of Salmonella compared to other pets. In people less than 21 years old in the USA, contact with reptiles or amphibians accounts for 11% of all sporadic Salmonella infections. These are therefore NOT recommended pets for any household with young children or individuals with a weakened immune system.

Normal, healthy pets should not be tested or treated for Salmonella, but Salmonella should be considered in animals that develop diarrhea. Transmission of Salmonella from a pet to a human in a household is very unlikely if appropriate precautions (as described above) are observed. Even if there are high-risk individuals in the household, diligent attention to infection control measures will minimize the risk of transmission. Given the well-described benefits of pet ownership, permanent removal of the pet is not indicated, unless extenuating circumstances exist which prevent proper infection control measures from being used. In these cases, the pet may be temporarily removed until its stops shedding Salmonella, but this would very rarely be warranted.

The risk of disease to the general population posed by Salmonella in house pets such as dogs and cats is:

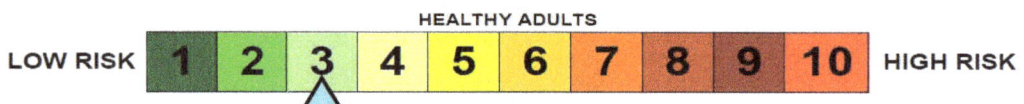

HEALTHY ADULTS

LOW RISK 1 2 3 4 5 6 7 8 9 10 HIGH RISK

Individuals with compromised immune systems (e.g. HIV/AIDS, transplant and cancer patients) are more susceptible to many kinds of infections, including those which may be transmitted by pets. While these individuals are not advised to get rid of their pets, precautions should be taken to reduce the frequency of contacts that could result in pathogen transmission (e.g. avoiding contact with any animal stool), as well as the ability of infectious agents to survive in the household (e.g. prompt and thorough disinfection of potentially contaminated surfaces).

Infants and young children (less than 5 years old) are more likely than adults to extensively handle animals if given the opportunity, more likely to touch their faces or mouths, and less likely to wash their hands after handling an animal. Children may "snuggle" with pets; this very close contact can increase the risk of disease transmission.

- Young children should be supervised when playing with animals, and an adult should ensure that they wash their hands afterwards, and especially prior to handling food. Older children should be taught to do the same.

- For these groups, the risk of disease posed by Salmonella in house pets such as dogs and cats is likely:

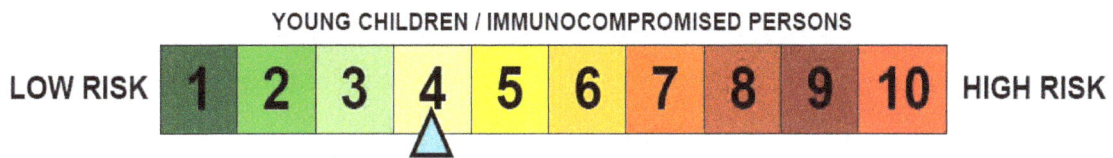

YOUNG CHILDREN / IMMUNOCOMPROMISED PERSONS

LOW RISK | 1 | 2 | 3 | 4 | 5 | 6 | 7 | 8 | 9 | 10 | HIGH RISK

Vibriosis

Vibriosis is a disease caused by an infection with bacteria of the Vibrio genus, most commonly Vibrio parahemolyticus or Vibrio vulnificus. Vibrio bacteria cause diarrhea, skin infections, and/or blood infections. The diarrhea-causing Vibrio parahemolyticus is a relatively harmless infection, but Vibrio vulnificus infection, though rare, can lead to blood poisoning and death in many cases.

Description

Vibriosis is a general term referring to an infection by any member of the large group of Vibrio, bacteria. The bacteria that causes cholera is in this group. Alternate names include non-cholera Vibrio infection, Vibrio parahemolyticus infection, and Vibrio vulnificus infection.

Vibrio parahemolyticus and Vibrio vulnificus are found in salt water. Infection with either of these two bacteria primarily occurs through eating contaminated raw seafood. Raw oysters are the usual source, although other seafood can carry the bacteria.

Vibrio parahemolyticus causes severe diarrhea. Vibrio vulnificus may cause diarrhea, but in persons with an underlying disease it may cause severe blood infections (septicemia or blood poisoning). Contact of a wound with seawater or contaminated seafood can lead to a Vibrio vulnificus skin infection.

Causes and Symptoms

Vibriosis is caused by eating seafood contaminated with Vibrio parahemolyticus or Vibrio vulnificus. These bacteria damage the inner wall of the intestine, which causes diarrhea and related symptoms. Vibrio vulnificus can get through the intestinal wall and into the bloodstream.

Persons at risk for severe, often fatal vibriosis include those with liver disease (cirrhosis), excess iron (hemochromatosis), thalassemia (a blood disorder), AIDS, diabetes, or those who are immunosuppressed.

Symptoms of intestinal infection occur within two days of eating contaminated seafood. Symptoms last for two to 10 days and include watery diarrhea, abdominal cramps, nausea, vomiting, headache, and possibly fever. Symptoms of a blood infection develop one to two days after eating contaminated seafood, and include fever, chills, low blood pressure, and large fluid-filled blisters on the arms or legs. Similar blisters can also be produced by a Vibrio vulnificus skin infection.

Diagnosis

Vibriosis can be diagnosed and treated by an infectious disease specialist. It is diagnosed when Vibrio bacteria are grown from samples of stool, blood, or blister fluid. The symptoms and a recent history of eating raw seafood are very important clues for diagnosis.

Treatment

To counteract the fluid loss resulting from diarrhea, the patient will receive fluids either by mouth or intravenously. Antibiotics are not helpful in treating Vibrio parahemolyticus diarrhea.

However, Vibrio vulnificus infections are treated with antibiotics such as tetracycline (Sumycin, Achromycin V), or doxycycline (Monodox) plus ceftazidime (Ceftaz, Fortraz, Tazicef). One out of five patients with vibriosis requires hospitalization.

Prognosis

Most healthy persons completely recover from diarrhea caused by Vibrio bacteria. Vibrio vulnificus blood infection affects persons with underlying illness and is fatal in half of those cases. Vibrio vulnificus wound infections are fatal in one quarter of the cases.

Prevention

Contamination with Vibrio bacteria does not change the look, smell, or taste of the seafood. Vibriosis can be prevented by avoiding raw or undercooked shellfish, keeping raw shellfish and its juices away from cooked foods, and avoiding contact of wounded skin with seawater or raw seafood.

Chlamydia

Chlamydia is one of the most common sexually transmitted diseases in the U.S. This infection is easily spread because it often causes no symptoms and may be unknowingly passed to sexual partners. In fact, about 75% of infections in women and 50% in men are without symptoms.

Infection with Chlamydia

It is not easy to tell if you are infected with chlamydia since symptoms are not always apparent.

But when they do occur, they are usually noticeable within one to three weeks of contact and can include the following:

Chlamydia Symptoms in Women

- Abnormal vaginal discharge that may have an odor
- Bleeding between periods
- Painful periods
- Abdominal pain with fever
- Pain when having sex
- Itching or burning in or around the vagina
- Pain when urinating

Chlamydia Symptoms in Men

- Small amounts of clear or cloudy discharge from the tip of the penis
- Painful urination
- Burning and itching around the opening of the penis
- Pain and swelling around the testicles

Diagnosis of Chlamydia

There are a few different tests your doctor can use to diagnose chlamydia. He or she will probably use a swab to take a sample from the urethra in men or from the cervix in women and then send the specimen to a laboratory to be analyzed. There are also other tests which check a urine sample for the presence of the bacteria.

Treatment of Chlamydia

If you have chlamydia, your doctor will prescribe oral antibiotics, usually azithromycin (Zithromax) or doxycycline. Your doctor will also recommend your partner(s) be treated to prevent reinfection and further spread of the disease.

With treatment, the infection should clear up in about a week or two. It is important to finish all of your antibiotics even if you feel better.

Women with severe chlamydia infection may require hospitalization, intravenous antibiotics (medicine given through a vein), and pain medicine.

After taking antibiotics, people should be re-tested after three months to be sure the infection is cured. This is particularly important if you are unsure that your partner(s) obtained treatment. But testing should still take place even if your partner has been treated. Do not have sex until you are sure both you and your partner no longer have the disease.

Chalmydia when not treated

If you do not get treated for chlamydia, you run the risk of several health problems.

- For women. If left untreated, chlamydia infection can cause pelvic inflammatory disease, which can lead to damage of the fallopian tubes (the tubes connecting the ovaries to the uterus) or even cause infertility (the inability to have children). Untreated chlamydia infection could also increase the risk of ectopic pregnancy (when the fertilized egg implants and develops outside the uterus.) Furthermore, chlamydia may cause premature births (giving birth too early) and the infection can be passed along from the mother to her child during childbirth, causing an eye infection, blindness, or pneumonia in the newborn.

- For men. Chlamydia can cause a condition called nongonococcal urethritis (NGU) an infection of the urethra (the tube by which men and women pass urine), epididymitis an infection of the epididymis (the tube that carries sperm away from the testes), or proctitis an inflammation of the rectum.

Prevention of a Chlamydia Infection

To reduce your risk of a chlamydia infection:

- Use condoms correctly every time you have sex.

- Limit the number of sex partners, and do not go back and forth between partners.

- Practice sexual abstinence, or limit sexual contact to one uninfected partner.

- If you think you are infected, avoid sexual contact and see a doctor.

Any genital symptoms such as discharge or burning during urination or an unusual sore or rash should be a signal to stop having sex and to consult a doctor immediately. If you are told you have chlamydia or any other sexually transmitted disease and receive treatment, you should notify all of your recent sex partners so that they can see a doctor and be treated.

Because chlamydia often occurs without symptoms, people who are infected may unknowingly infect their sex partners. Many doctors recommend that all persons who have more than one sex partner should be tested for chlamydia regularly, even in the absence of symptoms.

Gonorrhea

Gonorrhea is an infection caused by a sexually transmitted bacterium that can infect both males and females. Gonorrhea most often affects the urethra, rectum or throat. In females, gonorrhea can also infect the cervix.

Gonorrhea is most commonly spread during sex. But babies can be infected during childbirth if their mothers are infected. In babies, gonorrhea most commonly affects the eyes.

Gonorrhea is a common infection that, in many cases, causes no symptoms. You may not even know that you're infected. Abstaining from sex, using a condom if you do have sex and being in a mutually monogamous relationship are the best ways to prevent sexually transmitted infections.

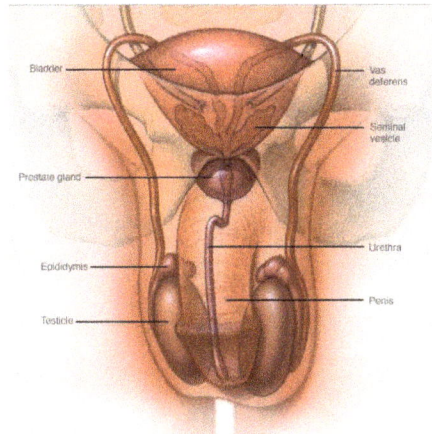

Male Reproductive System

Symptoms

In many cases, gonorrhea infection causes no symptoms. When symptoms do appear, gonorrhea infection can affect multiple sites in your body, but it commonly appears in the genital tract.

Gonorrhea Affecting the Genital Tract

Signs and symptoms of gonorrhea infection in men include:

- Painful urination

- Pus-like discharge from the tip of the penis

- Pain or swelling in one testicle

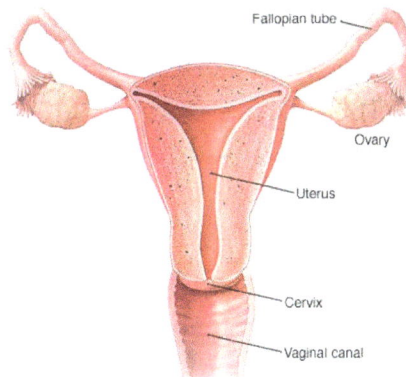

Signs and symptoms of gonorrhea infection in women include:

- Increased vaginal discharge

- Painful urination

- Vaginal bleeding between periods, such as after vaginal intercourse
- Painful intercourse
- Abdominal or pelvic pain

Gonorrhea at other Sites in the Body

Gonorrhea can also affect these parts of the body:

- Rectum. Signs and symptoms include anal itching, pus-like discharge from the rectum, spots of bright red blood on toilet tissue and having to strain during bowel movements.
- Eyes. Gonorrhea that affects your eyes may cause eye pain, sensitivity to light, and pus-like discharge from one or both eyes.
- Throat. Signs and symptoms of a throat infection may include a sore throat and swollen lymph nodes in the neck.
- Joints. If one or more joints become infected by bacteria (septic arthritis), the affected joints may be warm, red, swollen and extremely painful, especially when you move an affected joint.

When to See your Doctor

Make an appointment with your doctor if you notice any troubling signs or symptoms, such as a burning sensation when you urinate or a pus-like discharge from your penis, vagina or rectum.

Also make an appointment with your doctor if your partner has been diagnosed with gonorrhea. You may not experience signs or symptoms that prompt you to seek medical attention. But without treatment, you can reinfect your partner even after he or she has been treated for gonorrhea.

Causes

Gonorrhea is caused by the bacterium Neisseria gonorrhoeae. The gonorrhea bacteria are most often passed from one person to another during sexual contact, including oral, anal or vaginal intercourse.

Risk Factors

Factors that may increase your risk of gonorrhea infection include:

- Younger age
- A new sex partner
- A sex partner who has concurrent partners
- Multiple sex partners
- Previous gonorrhea diagnosis
- Having other sexually transmitted infections

Complications

Untreated gonorrhea can lead to significant complications, such as:

- Infertility in women. Untreated gonorrhea can spread into the uterus and fallopian tubes, causing pelvic inflammatory disease (PID), which may result in scarring of the tubes, greater risk of pregnancy complications and infertility. PID is a serious infection that requires immediate treatment.

- Infertility in men. Men with untreated gonorrhea can experience epididymitis — inflammation of a small, coiled tube in the rear portion of the testicles where the sperm ducts are located (epididymis). Epididymitis is treatable, but if left untreated, it may lead to infertility.

- Infection that spreads to the joints and other areas of your body. The bacterium that causes gonorrhea can spread through the bloodstream and infect other parts of your body, including your joints. Fever, rash, skin sores, joint pain, swelling and stiffness are possible results.

- Increased risk of HIV/AIDS. Having gonorrhea makes you more susceptible to infection with human immunodeficiency virus (HIV), the virus that leads to AIDS. People who have both gonorrhea and HIV are able to pass both diseases more readily to their partners.

- Complications in babies. Babies who contract gonorrhea from their mothers during birth can develop blindness, sores on the scalp and infections.

Prevention

Take steps to reduce your risk of gonorrhea:

- Use a condom if you choose to have sex. Abstaining from sex is the surest way to prevent gonorrhea. But if you choose to have sex, use a condom during any type of sexual contact, including anal sex, oral sex or vaginal sex.

- Ask your partner to be tested for sexually transmitted infections. Find out whether your partner has been tested for sexually transmitted infections, including gonorrhea. If not, ask whether he or she would be willing to be tested.

- Don't have sex with someone who has any unusual symptoms. If your partner has signs or symptoms of a sexually transmitted infection, such as burning during urination or a genital rash or sore, don't have sex with that person.

- Consider regular gonorrhea screening. Annual screening is recommended for all sexually active women less than 25 years of age and for older women at increased risk of infection, such as those who have a new sex partner, more than one sex partner, a sex partner with concurrent partners, or a sex partner who has a sexually transmitted infection.

Regular screening is also recommended for men who have sex with men, as well as their partners.

To avoid reinfection with gonorrhea, abstain from unprotected sex for seven days after you and your sex partner have completed treatment and after resolution of symptoms, if present.

Diagnosis

To determine whether the gonorrhea bacterium is present in your body, your doctor will analyze a sample of cells. Samples can be collected by:

- Urine test. This may help identify bacteria in your urethra.

- Swab of affected area. A swab of your throat, urethra, vagina or rectum may collect bacteria that can be identified in a laboratory.

For women, home test kits are available for gonorrhea. Home test kits include vaginal swabs for self-testing that are sent to a specified lab for testing. If you prefer, you can choose to be notified by email or text message when your results are ready. You may then view your results online or receive them by calling a toll-free hotline.

Testing for other Sexually Transmitted Infections

Your doctor may recommend tests for other sexually transmitted infections. Gonorrhea increases your risk of these infections, particularly chlamydia, which often accompanies gonorrhea. Testing for HIV also is recommended for anyone diagnosed with a sexually transmitted infection. Depending on your risk factors, tests for additional sexually transmitted infections could be beneficial as well.

Treatment

Gonorrhea Treatment in Adults

Adults with gonorrhea are treated with antibiotics. Due to emerging strains of drug-resistant Neisseria gonorrhoeae, the Centers for Disease Control and Prevention recommends that uncomplicated gonorrhea be treated only with the antibiotic ceftriaxone given as an injection in combination with either azithromycin (Zithromax, Zmax) or doxycycline (Monodox, Vibramycin, others) two antibiotics that are taken orally.

Some research indicates that oral gemifloxacin (Factive) or injectable gentamicin, combined with oral azithromycin, is highly successful in treating gonorrhea. This treatment may be helpful in treating people who are allergic to cephalosporin antibiotics, such as ceftriaxone.

Gonorrhea Treatment for Partners

Your partner also should undergo testing and treatment for gonorrhea, even if he or she has no signs or symptoms. Your partner receives the same treatment you do. Even if you've been treated for gonorrhea, you can be reinfected if your partner isn't treated.

Gonorrhea Treatment for Babies

Babies born to mothers with gonorrhea receive a medication in their eyes soon after birth to prevent infection. If an eye infection develops, babies can be treated with antibiotics.

Syphilis

Syphilis is an infection by the T. pallidum bacteria that is transmitted by direct contact with a syphilitic sore on the skin, and in mucous membranes.

A sore can occur on the vagina, anus, rectum, lips, and mouth.

It is most likely to spread during oral, anal, or vaginal sexual activity. Rarely, it can be passed on through kissing.

The first sign is a painless sore on the genitals, rectum, mouth, or skin surface. Some people do not notice the sore because it doesn't hurt.

These sores resolve on their own, but the bacteria remain in the body if not treated. The bacteria can remain dormant in the body for decades before returning to damage organs, including the brain.

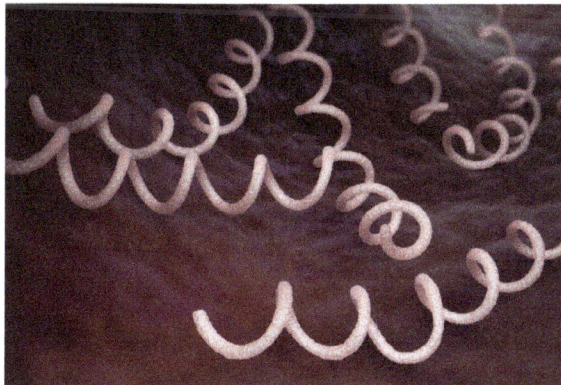

Syphilis is spread by the T. pallidum bacterium.

Causes

Syphilis is caused when T. pallidum transfers from one person to another during sexual activity.

It can also be passed from mother to a fetus during pregnancy, or to an infant during delivery. This is called congenital syphilis.

It cannot spread through shared contact with objects like doorknobs and toilet seats.

Risk Factors

Sexually active people are at risk of contracting syphilis.

Those most at risk include:

- Those who have unprotected sex
- Men who have sex with men
- Those with hiv
- People with numerous sexual partners

Syphilitic sores also increase the risk of contracting HIV.

Symptoms

- Syphilis is categorized by three stages with varied symptoms associated with each stage.

- However, in some cases, there can be no symptoms for several years.

- Contagious stages include primary, secondary, and, occasionally, the early latent phase.

- Tertiary syphilis is not contagious, but it has the most dangerous symptoms.

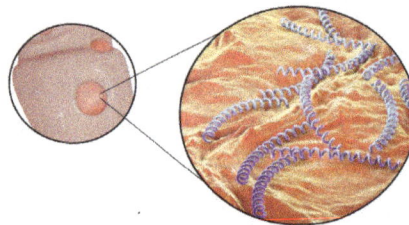

Syphilis is spread through the sores it causes, known as chancres.

Primary symptoms

The symptoms of primary syphilis are one or many painless, firm, and round syphilitic sores called chancres. These appear about 3 weeks after exposure.

Chancres disappear within 3 to 6 weeks, but, without treatment, the disease may progress to the next phase.

Secondary symptoms

Secondary syphilis symptoms include:

- a non-itchy rash that starts on the trunk and spreads to the entire body, including the palms of the hands and soles of the feet. It may be rough, red, or reddish-brown in color

- oral, anal, and genital wart-like sores

- muscle aches

- fever

- sore throat

- swollen lymph nodes

- patchy hair loss

- headaches

- weight loss

- fatigue

These symptoms can resolve a few weeks after they appear, or they can return several times over a longer period.

Untreated, secondary syphilis can progress to the latent and late stages.

Latent Syphilis

The latent phase can last several years. During this time the body will harbor the disease without symptoms.

After this, tertiary syphilis may develop, or the symptoms may never come back. However, the T. pallidum bacteria remain dormant in the body, and there is always a risk of recurrence.

Treatment is still recommended, even if symptoms are not present.

Late or Tertiary Syphilis

Tertiary syphilis can occur 10 to 30 years after onset of the infection, normally after a period of latency, where there are no symptoms.

Symptoms include:

- damage to the heart, blood vessels, liver, bones, and joints
- gummas, or soft tissue swellings that occur anywhere on the body

Organ damage means that tertiary syphilis can often be fatal.

Neurosyphilis

Neurosyphilis is a condition where the bacteria has spread to the nervous system. It is often associated with latent and tertiary syphilis, but it can appear at any time after the primary stage.

It may be asymptomatic for a long time, or it can appear gradually.

Symptoms include:

- dementia or altered mental status
- abnormal gait
- numbness in the extremities
- problems with concentration
- confusion
- headache or seizures
- vision problems or vision loss
- weakness

Congenital Syphilis

Congenital syphilis is severe and frequently life-threatening. Infection can transfer from a mother to her fetus through the placenta, and also during the birth process.

Data suggests that without screening and treatment, 70 percent of women with syphilis will have an adverse outcome in pregnancy.

Adverse outcomes include early fetal death, preterm or low birth weight, neonatal deaths, and infection in infants.

Symptoms in newborns include:

- saddle nose, in which the bridge of the nose is missing
- fever
- difficulty gaining weight
- a rash of the genitals, anus, and mouth
- small blisters on the hands and feet that change to a copper-colored rash and spread to the face, which can be bumpy or flat
- watery nasal fluid

Older infants and young children may experience:

- Hutchinson teeth, or abnormal, peg-shaped teeth
- bone pain
- vision loss
- hearing loss
- joint swelling
- saber shins, a bone problem in the lower legs
- scarring of the skin around the genitals, anus, and mouth
- gray patches around the outer vagina and anus

Tests and Diagnosis

A doctor will carry out a physical examination and ask about a patient's sexual history before carrying clinical tests to confirm syphilis.

Tests include:

- Blood tests: These can detect a current or past infection, as antibodies to the disease will be present for many years.
- Bodily fluid: Fluid from a chancre during the primary or secondary stages can be evaluated for the disease.

- Cerebrospinal fluid: This may be collected through a spinal tap and examined to test for any impact on the nervous system.

If there is a Diagnosis of Syphilis, any Sexual Partners Must be Notified of and Tested for the Disease:

Local services are available to notify sexual partners of their potential exposure to syphilis, to enable testing and, if necessary, treatment.

Healthcare providers also recommend HIV testing.

When to Get Tested

Many people will not know if they have an STI. It is a good idea to talk to a doctor or request a test:

- after having unprotected sex
- if you have a new sex partner
- if you have multiple sex partners
- if a sexual partner is diagnosed with syphilis
- if you are a man who has sex with different men
- if you have symptoms of syphilis

Anyone who is worried that they might have syphilis or another STI should speak to a doctor as soon as possible. Early treatment can cure it.

Treatment

Syphilis can be treated successfully in the early stages.

Early Treatment with Penicillin is Important, as Long-term Exposure to the Disease can Lead To-life-threatening Consequences:

During the primary, secondary, or late stages, patients will typically receive an intramuscular injection of Benzathine penicillin G.

The treatment strategy will depend on the symptoms and when the person was exposed.

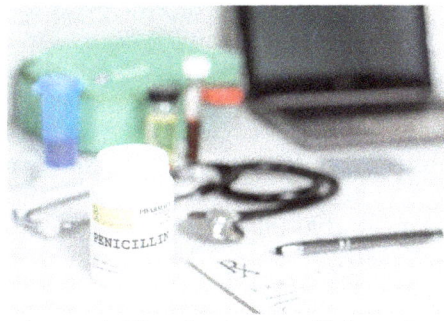

Syphilis can be treated using penicillin. The dosage will depend on the stage of the disease.

Tertiary syphilis will require multiple injections at weekly intervals.

Neurosyphilis requires intravenous penicillin every 4 hours for 2 weeks to remove the bacteremia from the central nervous system (CNS).

Curing the infection will prevent further damage to the body, and safe sexual practices can resume, but it cannot undo any damage that has already occurred.

Those with a penicillin allergy can sometimes use an alternative medication in the early stages. During pregnancy and in the tertiary stages, anyone with an allergy will be desensitized to penicillin to allow for treatment.

Following delivery, newborns who were exposed to syphilis in the womb should undergo antibiotic treatment.

Chills, fever, nausea, achy pain, and a headache may occur on the first day of treatment. This is referred to as a Jarisch-Herxheimer reaction. It does not indicate that the treatment should be stopped.

When is it Safe to have Sex

Sexual contact must be avoided until:

- all treatment has been completed.
- a blood test confirms that the disease has been cured.

It may take several months to see blood tests for syphilis go down to an appropriate level. This would provide confirmation of adequate treatment.

Prevention

Preventive measures to decrease the risk of syphilis, include:

- abstaining from sex.
- long-term mutual monogamy with an uninfected partner.
- condom use, although these protect only against genital sores and not those on the body. A variety of condoms are available to buy online.
- use of a dental dam, or plastic square, during oral sex. These are available for purchase online.
- not sharing sex toys.
- avoiding alcohol and drugs that could potentially lead to unsafe sexual practices.

Having syphilis once does not mean a person is protected from it. Once it is cured, it is possible to contract it again.

References

- Bacterial-disease, educational-magazines: encyclopedia.com, Retrieved 16 March 2018
- What-is-impetigo, health: healthline.com, Retrieved 19 April 2018
- Boils-skin-infection: onhealth.com, Retrieved 20 March 2018
- Campylobacter, fact-sheets: who.int, Retrieved 11 July 2018
- Vibriosis: medical-dictionary.thefreedictionary.com, Retrieved 15 April 2018
- Chlamydia, sexual-conditions: webmd.com, Retrieved 26 May 2018

Permissions

Index

www.ingramcontent.com/pod-product-compliance
Lightning Source LLC
Chambersburg PA
CBHW061259190326
41458CB00011B/3721